STEPPING INTO TIME

A Guide to Honolulu's Historic Landmarks

STEPPING INTO TIME

A Guide to Honolulu's Historic Landmarks

Text & Illustrations by
Jeannette Murray Peek

MUTUAL PUBLISHING

Copyright ©1994 by Mutual Publishing

No part of this book may be reproduced in any form or by any electronic or mechanical means including information storage and retrieval devices or systems without prior written permission from the publisher except that brief passages may be quoted for reviews.

All rights reserved

Design
Michael Horton Design

First Printing October 1994
1 2 3 4 5 6 7 8 9

ISBN 1-56647-024-2   Case
ISBN 1-56647-052-8   Soft

Mutual Publishing
1127 11th Avenue, Mezz. B
Honolulu, Hawaii 96816
Telephone (808) 732-1709
Fax (808) 734-4094

Printed in Taiwan

GLOSSARY OF HAWAIIAN TERMS

ali'i: Hawaiian chief, monarch or person of royal birth.
aloha: greeting, love, farewell.
haole: foreigner, especially Caucasian.
heiau: sacred burial ground.
hula: dance
kahili: feather standard, symbol of royalty.
kahuna: priest, minister, professional person.
kapu: taboo, prohibited, keep out.
koa: soldier, warrior; also, a native wood.
lauhala: pandanus leaf, especially as used in weaving.
luau: native Hawaiian feast.
makai: direction toward the sea.
mauka: direction toward the mountains.
mele: song or chant.

Table of Contents

Maps	vi
Acknowledgments	viii
Introduction by Glen Grant	ix
1. Alexander & Baldwin Building (SR, NR)	1
2. Aloha Tower (SR, HR)	4
3. Armed Services YMCA (NR)	6
4. Bernice Pauahi Bishop Museum (SR, NR)	9
5. C. Brewer Building (SR, NR)	12
6. Cathedral of Our Lady of Peace (SR, NR)	15
7. Diamond Head Lighthouse (NR)	17
8. Dillingham Transportation Building (NR)	19
9. ʻEwa Plantation	22
10. Falls of Clyde (NR)	24
11. Foster Botanic Garden (SR)	27
12. Hawaiian Electric Company Building (NR)	30
13. Honolulu Academy of Arts (NR)	32
14. Honolulu Hale (NR)	34
15. ʻIolani Palace (NHL)	36
16. ʻIolani Barracks (NR)	39
17. ʻIolani Palace's Coronation Pavilion (NR)	41
18. Kakaʻako Pumping Station (SR, NR)	43
19. Kamehameha V Post Office (SR, NR)	46
20. Kamehameha Schools (NR)	48
21. Kawaiahaʻo Church (NHL)	51
22. La Pietra	54
23. Linekona School (SR, NR)	57
24. Lunalilo Mausoleum (NHL)	60
25. McKinley High School	62
26. Mission Houses (NHL)	64
27. Moana Hotel (NR)	66
28. Oʻahu Railway & Land Company (NR, right-of-way; SR, cars)	69
29. Old Honolulu Police Station	72
30. Punahou School (NR)	74
31. Queen Emma Summer Palace (SR, NR)	77
32. Richards Street YWCA (NR)	80
33. Royal Brewery (NR)	83
34. Royal Hawaiian Hotel	85
35. Royal Mausoleum (SR, NR)	88
36. St. Andrew's Cathedral and Priory (NR)	90
37. Statue of King Kamehameha I and Aliʻiolani Hale (NR)	92
38. Territorial Building (NR)	96
39. Thomas Square (NR)	98
40. U.S. Post Office, Custom House and Court House (SR, NR)	101
41. USS *Arizona* Memorial (NHL, Pearl Harbor)	103
42. USS *Bowfin* (SR, NR, NHL)	105
43. War Memorial Natatorium (SR, NR)	108
44. Washington Place (NR)	111
Historical Notes	114
Bibliography	130

SR = STATE REGISTER OF HISTORIC PLACES
NR = NATIONAL REGISTER OF HISTORIC PLACES
NHL = NATIONAL HISTORIC LANDMARK

Map of Historical Sites

NUMERICAL LIST OF SITES

1. Alexander & Baldwin Building •
2. Aloha Tower •
3. Armed Services YMCA •
4. Bernice Pauahi Bishop Museum
5. C. Brewer Building •
6. Cathedral of Our Lady of Peace •
7. Diamond Head Lighthouse
8. Dillingham Transportation Building •
9. ʻEwa Plantation
10. Falls of Clyde •
11. Foster Botanical Garden •
12. Hawaiian Electric Company Building •
13. Honolulu Academy of Arts •
14. Honolulu Hale •
15. ʻIolani Palace •
16. ʻIolani Barracks •
17. ʻIolani Palace's Coronation Pavilion •
18. Kakaʻako Pumping Station •
19. Kamehameha V Post Office •
20. Kamehameha Schools
21. Kawaiahaʻo Church •
22. La Pietra
23. Linekona School •
24. Lunalilo Mausoleum •
25. McKinley High School •
26. Mission Houses •
27. Moana Hotel
28. Oʻahu Railway & Land Company
29. Old Honolulu Police Station •
30. Punahou School
31. Queen Emma Summer Palace •
32. Richards Street YWCA •
33. Royal Brewery •
34. Royal Hawaiian Hotel
35. Royal Mausoleum •
36. St. Andrew's Cathedral and Priory •
37. Statue of King Kamehameha and Aliʻiolani Hale •
38. Territorial Building •
39. Thomas Square •
40. U.S. Post Office, Custom House and Court House •
41. USS *Arizona* Memorial
42. USS *Bowfin*
43. War Memorial Natatorium
44. Washington Place •

• Sites with this symbol are located on enlarged downtown map

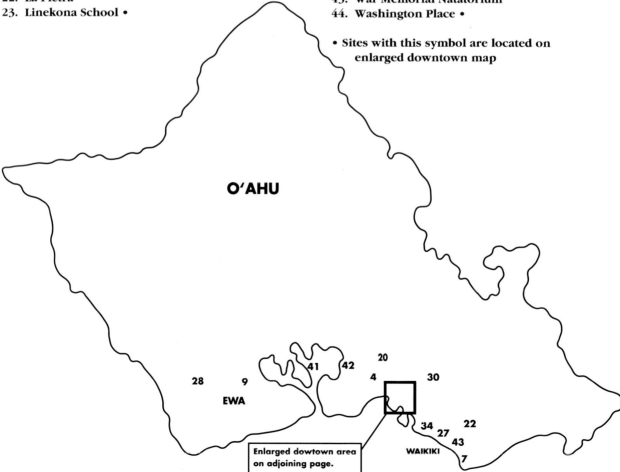

DOWNTOWN HONOLULU

MAP OF HISTORICAL SITES

Acknowledgments

This book evolved over the course of years, beginning as a collection of pen and ink drawings for which brief historic descriptions were provided. As my research broadened, I discovered the fascinating origins and events in history that gave meaning to these landmarks beyond their aesthetic appeal. Many people assisted me in the search for historic facts and architectural information. To all of the following I extend my deepest appreciation: Don Hibbard, Annie Griffin and Carol Ogata of the Hawai'i State Historic Preservation Division; Dion-Magrit Coshigano and Nettie Stilwell of the Historic Hawai'i Foundation; the staff of the State of Hawai'i Archives, Hawai'i State Library, University of Hawaii's Hamilton Library (Hawai'i/Pacific Collection), Honolulu Community College Library; Patty Belcher of the Bishop Museum Library; and Barbara Dunn and staff at the Hawaiian Historic Society; Janet Zisk, Archivist of the Kamehameha Schools; Mary Judd, Archivist of Punahou School; Paul Williams of the Hawaiian Railway Society; Mary Jo Valdes and Aldona Sendzikas at the USS *Bowfin* Museum; Father Paul Yim of the Catholic Diocese; and the staff and volunteers at Foster Gardens.

I also thank my sister, Laura Murray, for early reading of the manuscript and my husband, Joseph Peek, for assistance with word processing. My sincere appreciation to Glenn Grant for making history come alive through historical walking tours. Special thanks to Agnes Conrad for reviewing the manuscript for historical accuracy. Finally, my heartfelt gratitude to my parents and family for lifelong encouragement of my artistic pursuits. Mahalo nui loa!

INTRODUCTION
STEPPING INTO TIME: A GUIDE FOR TIME TRAVELERS

Jeannette Murray Peek and I first met in 1872 when she joined me and a group of fellow time travelers for a stroll down Merchant Street past the King Kamehameha V Post Office, Mr. Charles Reed Bishop's new bank, and the government offices of Honolulu Hale before stopping for an "Old-Fashioned" at E. S. Cunha's Union Saloon. A few weeks later we met again at Ali'iolani Hale in the revolution-torn days of 1893 — that was right after we had journeyed to the 1831 village of Honoruru to visit with the American Protestant missionaries at their Kawaiaha'o compound. Time traveling with Jeannette may have become commonplace, but it was never dull.

Most people mistakenly believe that stepping into time requires an H. G. Wells machine or a Hollywood nuclearized DeLorean. Actually, the knack of time travel, as any aficionado of the fantastic sport can tell you, is wholly dependent upon a flexible imagination, a burning desire to cut the threads of the present, and a physical place where the sense of history has survived. In his illustrated novel *Time and Again*, author Jack Finney observes that in the modern American city, in spite of all the indistinguishable glass and steel monstrosities, "there are fragments of still earlier days" in older buildings which are "still-surviving fragments of a clear April morning of 1871, a gray winter afternoon of 1840, a rainy dawn of 1793." If one were to surround him or herself in that historic setting, recreating the clothing, foods, habits, language and paraphernalia of a former time, then it may be possible, Finney suggests, for that person to step out of *now* and plunge headlong into *yesterday*.[1]

For the Hawai'i time traveler, Jeannette Murray Peek has created, with *Stepping Into Time*, the perfect guide for finding the ideal historic landmark to use as a launching pad to past eras. While I have attempted through costume, story and plenty of imagination to recreate through walking tours a past that has been displaced by wrecking balls, Jeannette has used the artist's brush and writer's pen to recreate an historic Honolulu surviving amidst the glass and steel boxes. Through her talented drawings and informative text, you will have the opportunity to attend an elegant soiree in royal Nu'uanu Valley at Hanaiakamalama, the summer palace of Queen Emma, or relive the plantation days of Ewa, or check into the Armed Forces YMCA in war-braced Honolulu. The train station of the Oahu Railway and Land Co., the red-brick eminence of the Royal Brewery, and the stylish facade of the Honolulu Police Station—long ago converted for other uses — will once again open their doors to the gusto of bustling passengers, Primo beer and hardened detectives. Within this time travel guide are the remaining ghosts of a Honolulu which existed on the morning of December 7, 1941, that witnessed the overthrow of the Hawaiian monarchy on January 17, 1893, or that set the backdrop for the coronation ceremonies of King Kalakaua in February 1883. History cannot be more vivid.

As a community, we are frequently asked, "What is the value of our old buildings?" Time travelers, of course, have a quick answer — without the elegance of an Alexander and Baldwin Building or the uniqueness of the Kaka'ako Pumping Station, we have nothing left but old photographs, drawings or the imagination to reconstruct the past.

For others ensnared in the threads of the present and looking only to the future, older structures are more often inconvenient rattraps that are better replaced with more functional high-rises. One need only view the skyline of downtown Honolulu or Waikiki to know in an instant the attitude most developers have had concerning what would be best for our built environment.

In that sense, reading *Stepping Into Time* is not only nostalgic, but painful. When Samuel "Mark Twain" Clemens visited Honolulu in 1866, the great American writer was duly impressed. "The further I traveled through the town," he noted, "the better I liked it. Every step revealed a new contrast — disclosed something I was unaccustomed to."[2] By the 1930's, the emergence of an "Hawaiian architectural" style associated with architects such as Lincoln Rogers, O. G. Traphagen and C. W. Dickey had reinforced Honolulu's reputation as being distinctly unlike any of its urban cousins to the East or West. In the pell-mell rush to modernity, that island distinctiveness seems only to remain in our historic buildings and in our revived interest in discovering these fading remnants of yesteryear.

A city without character is in danger of losing its soul. Time traveling is not only pleasurable, but vital to recapture a spirit of pride in Hawai'i's past and thereby spark a commitment to develop the future with a respect for the multicultural uniqueness of our built environment. As a community, we need to listen to the stories that the old decaying relics in Mo'ili'ili and Kalihi or neglected small towns along the Hamakua Coast tell us about who we are as an island society and what is valuable to pass on through the generations. The fast-food convenience architecture that is quickly dominating our neighborhoods can't tell us any stories except those transplanted from elsewhere. Stepping into time is essential if we are to create a meaningful tomorrow.

Glen Grant
Honolulu TimeWalks, Mo'ili'ili, Hawaii

[1] Jack Finney, *Time and Again: An Illustrated Novel*, (N.Y.: Simon and Shuster, 1970), pp. 56.
[2] Samuel Clemens, *Roughing It in the Sandwich Islands* with a Foreword by A. Grove Day, (Honolulu: Mutual Publishing Co., 1990), p. 1.

Alexander & Baldwin Building

The Alexander & Baldwin Building was the third headquarters built for the corporation founded by Samuel Thomas Alexander and Henry Perrine Baldwin. The previous headquarters had been housed in the Judd Building and the Stangenwald Building. Located in the heart of Honolulu's financial and business district, it occupies an 18,000-square-foot lot and was the first business house that could be considered a truly Hawaiian building. It led Hawai'i away from copying other trends to develop its own regional style.

When completed in 1929, the Alexander & Baldwin Building was considered the finest structure in Honolulu, blending many architectural elements to form a uniquely decorative, yet functional, edifice. It was also fireproof, using no woodwork in its construction or finish. After purchasing successive neighboring lots, Alexander & Baldwin had acquired the entire frontage on Bishop Street from Merchant to Queen streets.

From the beginning, the company was very much a family affair. Wallace Alexander, Harry Baldwin, John Waterhouse, and other family members took an active interest in the planning and construction of the edifice. The chief architect, C. W. Dickey, was a relative of both the Alexanders and the Baldwins. He popularized the now famous "Dickey roof," characterized by a high peak and wide overhangs, which he first used in residential architecture. Dickey's junior partner on the project was Hart Wood, one of the first Western architects to introduce Oriental styles to Hawai'i. Wood used a wide variety of architectural styles and was a master of ornamentation, combining various forms and influences into a harmonious whole. While Dickey prepared the general design and oversaw the project, Wood was responsible for most of the decorative detail.

The building includes many Chinese features, as well as Mediterranean, Italian Renaissance, Buddhist, Tibetan, Japanese, and Hawaiian elements and ornamentation. Its portico is a blend of cultural influences, with Hawaiian murals, Chinese good-luck symbols, Japanese grillwork on the doors, and even Egyptian fluting on the columns textured to resemble papyrus. At a cost of around $1,240,000, the structure was an enlightened, and costly, accomplishment for the time. Its successful integration of diverse architectural styles and features received instant acclaim. Dickey explained that his foremost thought was to create a structure suitable to the climate, environment, history, and geographic position of Hawai'i.

The early history of the sugar industry upon which the Alexander & Baldwin firm was founded was closely linked with Chinese labor. Hawai'i was considered the crossroads of the Pacific, and its proximity to the Orient gave sufficient reason for allowing Chinese architecture to influence the design. Dickey did not, however, want any part of the building to look as though it had been transplanted from China, so the final result shows only subtle Chinese influences. On the exterior, this influence is found in the water buffalo heads, the quaint Chinese faces of the window ornamentation, the good-luck signs on the main entrance portico on Bishop Street, and long-life signs on the column capitals. Chinese motifs appear again in the tilework of the main entrance portico surrounding the panels of Hawaiian fishlife. Pictorial tile decorations, especially designed for the public room, are in the form of two large ceramic tile murals placed 29 feet above the floor. One mural, painted by California artist Jessie Stanton, depicts Maui's 'Iao Valley and Kahului Harbor. The other, by Einar Peterson, shows the sailing ship *John Ena* at Port Allen, Kaua'i.

Front and rear exterior walls are of matte-glazed terra cotta tied to the concrete backing by copper

ALEXANDER & BALDWIN BUILDING
HONOLULU, HAWAII

anchors. The roof is of steel and reinforced concrete covered with a built-up felt and asphalt roof over which is laid the colored roofing tile. A tropical setting was created for the building by landscape architect Richard C. Tongg. Cypress grass and full-grown coconut trees were added around the exterior. The trees created quite a stir in Honolulu as "the palm trees that grew overnight." The sidewalk was painted green to absorb the glare, make the area cooler, and blend in with the landscape.

In recent years, the Alexander & Baldwin Building has undergone major remodeling, but utilization of the structure has essentially remained the same. The lobby once had 39-foot ceilings with five bronze chandeliers, each with thirty-five globes. In a 1955 renovation, a floating mezzanine lowered the ceiling and the chandeliers were removed. One of the tile wall murals is now in the coffee shop and the other is in the boardroom. Vinyl coating and walnut paneling were added on some walls, but many of the original travertine walls remain.

• *See Historical Notes (page 114) for more information.*

The Alexander & Baldwin Building, located at 822 Bishop Street in downtown Honolulu, was built as the corporate headquarters and monument to the founders of Alexander & Baldwin, Inc., one of Hawaii's "Big Five" corporations which dominated economic policy and politics until World War II. Its style is a unique blend of Oriental and Western architecture.

Aloha Tower

ONE OF THE MOST RECOGNIZABLE LANDmarks of Hawai'i is Aloha Tower, which stands at the foot of Fort Street, overlooking Honolulu Harbor. So prominent a feature of the Honolulu waterfront, it has been called the "Liberty Statue of the Pacific." At the time of its completion in 1926, it was the tallest building in Honolulu and the pride of the islands.

By the 1900s, sea trade in Honolulu had grown to such proportions and at such a rapid pace that traffic in Honolulu Harbor was a constant problem. The demand for sugar and pineapple, of which Hawai'i was the leading producer, accounted for much of the growth in trade. The first measure taken to improve conditions for sea-going vessels was to widen and dredge the harbor and to form the Honolulu Port Authority to study the situation. In 1920 the Port Authority initiated plans to build a tower to house its offices and the harbor master. The plan was to create not just an office building but a unique structure to welcome all visitors and seafarers—a beacon on the skyline for tourists and pilots. The 184-foot tower was designed by architect Arnold Reynolds and built at a cost of $190,000. Its design included four huge clocks, each weighing seven tons, one on each of its four sides. These were shipped from the Howard Clock Company in Boston and were guaranteed to be accurate to within thirty seconds per month. The architectural style of the tower could be described as modernistic Gothic. Above its dome is a mast with a crossarm at the top and cables used to hang signals for transmitting information to ships in or near the harbor. On the tenth floor is an observation balcony that originally housed a lookout station for the harbor pilots. The word "ALOHA" is inscribed on all four sides of the tower.

In 1925 the Honolulu newspapers headlined "Aloha Tower, Gateway to Fort Street, Completed." In 1926 a 5,500-candlepower light was installed atop the green-colored dome. The light was visible for 19 miles at sea as it traveled through an arc of 180 degrees.

Once dominating the Honolulu city skyline, Aloha Tower is dwarfed by its neighbors today. Still, from its tenth-floor observation deck one can enjoy an unobstructed panoramic view of Honolulu Harbor.

• *See Historical Notes (page 114) for more information.*

Once the tallest building on the Honolulu waterfront, Aloha Tower now seems miniscule amid the towering skyscapers of downtown Honolulu. Located at Pier 9, this well-known landmark is open daily, 8:00 AM–8:00 PM. It features a museum on the ninth floor and a panoramic view of Honolulu from the tenth floor.

Armed Services YMCA

JUST ONE BLOCK *MAUKA* OF ʻIOLANI Palace on the Richards Street side stands an impressive Spanish Mission style stucco building, elegant in design with its cast stone detailing, iron grillwork, and ornamental light fixtures. Surrounded by a wrought iron fence almost identical to that around the palace grounds, it appears much like a palace itself. Originally the home of the Armed Services YMCA, the run-down building was purchased in 1987 and renovated to serve as the headquarters for the Hemmeter Corporation. A massive $30 million reconstruction project financed by the building's owner, Chris Hemmeter, transformed the old YMCA and won for it the Historic Building Award in 1989. Hemmeter later sold the building to the BIGI Corporation for $80 million. Impressive for its time, the early YMCA building was almost, but not quite, as elegant as the building we see today.

The first Army and Navy YMCA, as it was originally called, was opened on November 25, 1917, in response to a need for a large facility for the increasing military population in Honolulu. Government officials had argued the special need for such a facility for two reasons: the age limit for enlistment had been reduced from eighteen to sixteen years, and recruiting was now being done throughout the interior states, resulting in a "higher type" of young serviceman. Furthermore, the number of men stationed on Oʻahu during World War I was expected to increase from 9,000 to 25,000. There was no place in Honolulu where an enlisted man could obtain sleeping quarters at a reasonable price and there

This site, at 1 Capitol District on the corner of Richards and Hotel streets, is renowned for both the original Royal Hawaiian Hotel and the Armed Services YMCA. Today this building, known as the Hemmeter Corporation Building, houses private corporate and state offices.

7

ARMED SERVICES YMCA

was virtually no organized entertainment. The notorious red light district of Honolulu was thriving by this time, and many local citizens feared for the moral welfare of the young enlisted men. Providing them constructive religious work and healthy recreation was considered to be of the utmost importance.

The YMCA set its sights on the old Royal Hawaiian Hotel on Hotel Street between Richards and Alakea in the block beyond the Central YMCA. It received generous donations from the Alexander Young estate and raised the astonishing sum of $250,000 to purchase the hotel. This "handsomest building of tropical style in the city" had been the leading hotel in Honolulu for forty-five years and held the distinction of having housed foreign visitors and friends of the royal family. Little change was needed other than to remodel the first floor and rebuild one cottage for a gymnasium and another for a dormitory.

By 1924, attendance at the "Y" was equal to eight times the total population of Honolulu. Every available bit of space was converted to sleeping quarters as more and more men, and then their families, required accommodations. By 1926, age, termites, and heavy use had taken their toll on the building and it was torn down to be replaced by what was called the "largest and most complete" building of its kind in the world. The architectural firm of Emory and Webb, and particularly San Diego architect Lincoln Webb, are credited with the Latin design of the YMCA building. It complemented the Federal Building, the Hawaiian Electric Company building, Honolulu Hale, and the new Royal Hawaiian Hotel at Waikiki, all built around the same time. Like the building it replaced, the concrete and stucco building was U-shaped, open and airy, with large lanais. Its central court contained a swimming pool, and its spacious and well-landscaped grounds were mostly preserved in their original state. It was five stories high, with a fifth-floor open loggia. The central lobby offered a billiard room on the right and a cafeteria on the left, with a gym and offices behind. It also featured locker rooms, a barbershop, a tailor, and a curio shop. The second floor contained more game rooms, a writing room, reading lanai, auditorium, and meeting rooms. A color scheme of blue and sand was used throughout. The Italian scrollwork and other decorations were modeled after the Davanzatti Palace in Florence, Italy. The third, fourth, and fifth floors contained 268 sleeping rooms. Room rates were twenty-five cents per night in the dormitory and one dollar per night for a room with a bath. During World War II, the Army and Navy YMCA was renamed the USO Army and Navy Club. It afforded servicemen a centrally located place where classes, activities, dances, silent movies, and auctions were alternatives to other activities available just a short distance away on Hotel Street and across the street at the infamous old Black Cat Cafe. When the Pacific Fleet paid its occasional visits, the Army and Navy "Y" would come alive with activity. Honolulu, then a city of 100,000, rallied to the cause of entertaining 45,000 sailors for fifty days.

The Army and Navy Club, renamed the Armed Services YMCA after World War II, played a major role in the lives of U.S. military personnel throughout the war years. On its tenth anniversary in 1938, it was still gaining recognition as one of the finest centers in the world.

Bernice Pauahi Bishop Museum

PRINCESS BERNICE PAUAHI WAS ONE of the first pupils of the Royal School, established by the missionaries at the request of King Kamehameha III. As the last direct descendant of the Kamehameha dynasty, the heiress to family property and wealth was noted for her philanthropy and selfless dedication to the Hawaiian people. In 1846 she was introduced to the newly arrived Charles Reed Bishop, formerly of Glens Falls, New York. Bishop was en route to California, but after seeing Hawai'i he decided to make it his home. Bernice Pauahi and Bishop were married in 1850. He went on to become Honolulu's first banker. During their married life, Mrs. Bishop had expressed a desire for a proper place in which to display all her inherited treasures. She died in 1884, and five years later her husband founded the Bernice Pauahi Bishop Museum of Polynesian Ethnology and Natural History as a memorial to her.

The original Hawaiian National Museum had been housed in an upstairs room of Ali'iolani Hale. It lacked a resident curator and experienced personnel and was criticized by the community for its failure to provide a worthy facility dedicated to the culture and traditions of Hawai'i and the Pacific. The impetus to create a separate museum grew out of this early attempt at a Hawaiian museum. The site chosen for the new museum was the campus of the newly established Kamehameha Schools, which Princess Bernice had provided for in her will. Construction on the original museum building (Museum Hall) began in 1888 and was completed in the summer of 1890. It opened to the public in June 1891. Bishop Hall was constructed at the same time, but was finished slightly later; it opened as part of the Kamehameha School for Boys and was formally dedicated on December 19, 1891. Bishop Hall was transferred to Bishop Museum in 1961, several years after the Kamehameha Schools moved to their present location.

The first museum building, Museum Hall, was designed in a Romanesque Revival style by William F. Smith of the San Francisco firm Smith and Freeman. He used local lava rock, called Kamehameha blue-stone, for the exterior and koa wood for the interior. Museum Hall consisted of the main entrance and tower flanked by a one-story exhibit room, a two-story exhibit area, and a picture gallery. Polynesian Hall was completed in 1894, although it opened to the public on a limited basis before its completion in June 1891. Hawaiian Hall was completed in September 1900, but did not open until November 24, 1903. Paki Hall (1911), Konia Hall (1926), and Castle Hall (1989) were later added to accommodate the museum's growing collection and special exhibits.

Bishop Museum, recently named the State Museum of Natural and Cultural History, is a world center for the study of Pacific Island culture and for marine research. Its galleries and planetarium offer instructional displays of artifacts, technology, and customs from throughout Polynesia, Micronesia, and Melanesia. Its Hawaiian collection, the only one of its kind in the world, includes gourd and shell utensils, ceremonial carvings, and rare featherwork and barkcloth—the remnants of crafts that thrived generations ago in Hawai'i but have almost disappeared in modern times.

• *See Historical Notes (page 114) for more information.*

Bishop Museum, located at 1525 Bernice Street in the Kalihi district, features a collection of Hawaiian and Polynesian artifacts and crafts, planetarium, museum shop, and restaurant. Lectures and special traveling exhibits are among its cultural and educational offerings. The museum is open daily, 9:00 AM–5:00 PM. Telephone (808) 847-3511 for information.

C. Brewer Building

A RICH HISTORICAL FOUNDATION precedes the C. Brewer Building and its association with the oldest American business firm west of the Rockies—C. Brewer and Company. Captain James Hunnewell, first officer of the brig *Thaddeus* which brought the first missionaries to Hawai'i, bought the premises in 1826 for the sum of $250. He began operating a trade and shipping business of sandalwood, cattle hides, and goat skins from Hawai'i, merchandise from New England, and tea and spices from China.

When Captain Charles Brewer became a partner in 1836, the business was being run by Henry A. Peirce and was furnishing supplies for the booming North Pacific whaling industry. Brewer (1804–1885) was born in Boston and by age twenty-five was the first officer on the brig *Chinchilla*, based in Honolulu. Peirce and Brewer began trading cotton, rum, and American-made goods in return for sandalwood and furs. When Peirce returned to New England in 1843, Brewer took full control of the company, which operated under his name. Foreseeing the decline of the whaling industry in the 1850s, the company focused on sugar and soon established itself as a leader in sugar production. Later, to diversify, it entered the macadamia nut industry in 1959 and soon grew to be one of the world's major producers.

The first C. Brewer Building was a two-story brick structure on Queen Street designed by Clifton Briggs Ripley and Charles W. Dickey in 1898. By 1925, through consolidation with other companies, C. Brewer

Distinguished by its diminutive size and homelike appearance, the C. Brewer Building, located at 827 Fort Street in downtown Honolulu, is the oldest firm still active in Hawai'i. It is an outstanding Hawaiian adaptation of the Spanish Mission Revival style of architecture.

13

C BREWER BUILDING

& Company was one of Hawai'i's largest corporate landholders and controlled 25 percent of the island's sugar—about one million tons annually.

In the late 1920s, the company decided it needed a new corporate headquarters and hired architect Hardie Phillip for the project. Phillip designed the new C. Brewer Building in 1930 with a character that was distinctly Hawaiian and homelike. The two-story building, set amid a tropical garden, used elements that had been successfully employed in the Academy of Arts building—a combination Mediterranean style, with a Hawaiian motif incorporating the high double-pitched tile hip roof with widespread overhang referred to as the "Dickey roof" after the architect who first introduced it. For the exterior, Phillip chose a combination of concrete and native Hawaiian blue-stone with a textured stucco finish and natural gray color. The projecting second-floor lanai of wrought iron and concrete is decorated with Hawaiian motifs. Inside the main entrance, beyond the polished koa wood doors, are decorative wrought iron railings and grillwork representing windswept sugar cane. Modern light fixtures were designed to resemble sugar cubes. Interior finishes are of ohia and teakwood, and floors and walls are of island sandstone. Upstairs offices open onto balconies overlooking gardens, and ground-floor offices open into the gardens and courtyards. An overall feeling of coolness and comfort pervades the building.

The architectural influences in the construction of the C. Brewer Building are Spanish, Mediterranean, and Hawaiian. It is enclosed by a stucco-covered wall broken by occasional grills of stylish cast iron. It has garages in back, and perched atop the building is the old Hudson Bay Company's beaver weathervane. Though originally housing only the company's upper echelon, this is the last and smallest headquarters to be built for one of Hawai'i's "Big Five" corporations. Today, the C. Brewer Building continues to hold a place of prominence on Fort Street amid the downtown skyscrapers—a charming island in itself recalling earlier days of commerce in a smaller, quieter Honolulu.

• *See Historical Notes (page 114) for more information.*

Cathedral of Our Lady of Peace

The first group of Catholic missionaries to Hawai'i, led by Father Alexis Bachelot, sailed from France in November 1826 aboard *La Comete* and arrived in Honolulu on July 7, 1827. Ka'ahumanu, the queen regent and staunch supporter of the Protestant missionaries, ordered them to leave immediately, but the captain who had brought them to the islands paid no heed and set sail without them. For about a year there were no serious objections to their presence, and the priests kept to themselves on the land Philippe de Morineau, leader of the French colony in Honolulu, had secured for them from King Kamehameha III in 1827. There the priests—Fathers Bachelot, Short, and Armand—and three lay brothers built their first chapel, cultivated the land, and prepared for their missionary work by learning the Hawaiian language.

By 1829 the priests had baptized sixty-five adults and some children and were teaching the catechism. Alarmed by this small show of progress, the Protestant missionaries began to express their opposition. A period of persecution and banishment of the priests followed. It was not until 1839, when King Kamehameha III passed the Edict of Toleration allowing freedom of religion, that the Catholic mission was able to continue its work.

Bishop Rouchouze arrived in May 1840, bringing Father Maigret with him, and soon after their arrival he contracted F. J. Greenway to build a small coral block church. After laying the cornerstone, the bishop returned to France to recruit more missionaries and obtain supplies for the mission; on the return voyage, his ship was lost at sea. Father Maigret, who had been in charge during the bishop's absence, was then appointed bishop and became the force behind the establishment of the Catholic mission. Bishop Maigret supervised the construction of the church that would be called Our Lady of Peace, "Notre Dame de la Paix."

The simple Roman-style church was built of coral blocks on the mission site established in 1828. It had a rustic wooden altar, plain wooden benches, and a floor covered with *lauhala* matting. The first pipe organ to be brought to Hawai'i was shipped from France and installed in 1846. Hawaiians were thrilled with the lovely music that resounded inside the church. No doubt the organ brought many a curious native to its doors. Two bells for the tower were also imported from France, and its Parisian clock—the oldest tower clock in Hawai'i—has been in constant operation since about 1846. The style of the church later became more Gothic, and between 1870 and 1880 major improvements took place, including the addition of stained glass windows and statuary imported from France. From its simple origins, the coral church evolved into an attractive cathedral, with a new concrete porch added in 1910 and the original belfry replaced by a concrete tower in 1917. Under Bishop Alencastre, all interior woodwork was replaced in 1926, and in 1927 the congregation donated a new altar and two white marble statues of Our Lady and Saint Joseph. The Doric columns in front were added in 1929, and Maigret's original windows were replaced with fourteen stained glass windows imported from Germany.

A kiawe tree planted by Father Bachelot is the ancestor of all Hawai'i's kiawe trees and was the last survivor of plants grown by Father Bachelot from seeds he had brought from the royal gardens in Paris. It stood for ninety-one years and is a symbol of the Catholic church in Hawai'i. The stump of the old tree is preserved in the courtyard beside the cathedral.

Our Lady of Peace Cathedral is the oldest Catholic church in the islands and the oldest Catholic cathedral in continuous use in the United States. Next to the cathedral on the *mauka* side is a statue of Our Lady of Peace, whose four-sided pedestal is inscribed in Hawaiian, English, French, and Portuguese with, "In memory of the First Roman Catholic Church. Our Lady of Peace. 1827–1893."

• *See Historical Notes (page 115) for more information.*

Our Lady of Peace Cathedral, located at 1183 Fort Street on the Beretania side of the Fort Street mall, was the first Catholic cathedral built in Hawai'i. Its grounds were also the site of the first Catholic mission to the Hawaiian Islands. The cathedral continues to serve a large congregation.

Diamond Head Lighthouse

With a burgeoning sea trade between Hawai'i and the U.S. mainland and Pacific Rim countries in the late 1800s, the need to assist vessels safely into Honolulu Harbor became paramount. Before lighthouses were erected in the islands, young boys announced arriving ships by running through the town blowing conch shells. In 1856 a simple wooden structure was built as a lookout tower for ships approaching the harbor. That same year, a telegraph station was constructed on Punchbowl Hill from which signals were sent to the town below to announce the approach of vessels. A lookout station was also situated on Diamond Head in 1878, and ten years later the Barber's Point Lighthouse was built.

The lower slopes of Diamond Head crater were considered a prime location from which to guide mariners. In 1897 construction began on the Diamond Head Lighthouse, and it was completed in 1899. The initial structure, a four-sided, whitewashed stone tower with a black lantern, was replaced in 1918 by a 55-foot reinforced concrete tower built in the classic lighthouse design commonly found in coastal regions of the United States. Its third-order Fresnel lens and 7,300-candlepower lamp cast light that could be seen 18 nautical miles. The lighthouse was operated by the Lighthouse Service. The original lens is still in place today, but the system is now fully automatic and uses an electric lamp with 50,000 candlepower for the white section and 11,000 candlepower for the red section. Next to the Diamond Head Lighthouse is a house and a guest cottage which originally served as the residence of the superintendent of the 19th Lighthouse District. In 1939 the Lighthouse Service merged with the U.S. Coast Guard, and since then the lighthouse has served as quarters for the Coast Guard's 14th District commander. It was also the radio station for this Coast Guard district until 1945. This enviable three-acre site, perched 147 feet above the ocean, provides an unparalleled view and ideal vantage point for the lighthouse.

The Diamond Head Lighthouse is one of nine lighthouses in the Hawaiian Islands and one of the best known in the world. It serves as the marker for the end of the Transpac Yacht Race. The commander's residence is designed as a one-story bungalow with a gabled roof, exposed rafters, and overhanging eaves. It sits on a raised rock column foundation, with latticework between the columns. The lot is enclosed on the east and south by a wall of coral, blue-stone, and moss rock. Though its function is less important today, the Diamond Head Lighthouse is still an appealing sight on the slopes of O'ahu's most familiar landmark and a reminder of one of the earliest attempts to modernize navigation in the islands.

On the slopes of Diamond Head, at 3399 Diamond Head Road, stands one of the best known beacon lights in the Pacific. The 55-foot-high tower is operated by the U.S. Coast Guard. Next to the lighthouse is the residence of the Coast Guard's 14th District commander.

Dillingham Transportation Building

Walter F. Dillingham (1875–1963) was the son of Benjamin Franklin Dillingham, the founder of Hawai'i's first railway system and one of the developers of 'Ewa Plantation. The elder Dillingham had come to Hawai'i from West Brewster, Massachusetts, as first officer of the bark *Whistler*. In 1864 he arrived in Honolulu, where he was detained by a broken leg, and, after a long recuperation, he decided to remain. He first found employment in a hardware store, which he later bought and named Dillingham & Company. The enterprise grew to the extent that Dillingham became one of the prominent figures in the railway and sugar industries in Hawai'i. His son Walter followed in his footsteps and extended his business interests to the neighbor islands. He promoted the building of a headquarters in Honolulu which would bear the Dillingham name and serve as a tribute to his father.

Dillingham Transportation Building was designed in a Mediterranean style, with Italian Renaissance features by architect Lincoln Rogers of Los Angeles. It was completed in 1929. Another of Rogers' well-known works is the Army and Navy YMCA on Hotel Street (see Armed Services YMCA). The formal opening ceremonies in September 1930 brought many an admirer to see the colossal four-story, block-long building. Its archways, framing an arcade with beautifully painted eaves and medallions displaying sailing vessels and the face of a rugged sailor, are part of the ornamentation that won acclaim. Three connected wings are tied together visually by the arcade that extends across the front. The first floor was built of cut stone and is separated from the upper floors by a classical cornice. The ground floor contains shops and stores, and the upper floors were designed specifically for offices. Other unique elements are the central ground-floor lobby with walls of Spanish marble and a floor with multipatterned tiles representing a ship's compass and swirling waves of the sea. The ceiling is done in green, red, black, and gold designs with paneled beams.

Children's Hospital acquired the stock of the building in 1945 and took over management soon after. In the early 1970s the site was purchased on a long-term lease by Grosvenor International, who considered demolishing the low-rise building to make way for more profitable commercial high-rises. Fortunately, a decision was made to develop the surrounding area instead—to create two silver-mirrored office buildings thirty stories high and to restore the Dillingham Transportation Building to its original form. The result was a combination of old and new, with the beautiful historic facade of the old building reflected in the shiny glass towers of Grosvenor Center. Architects Hawai'i, the firm in charge of the project, received well-deserved praise for an accomplishment that served to restore and integrate one of the last outposts of Bishop Street's "gateway to the harbor" into the modern Honolulu financial district.

• *See Historical Notes (page 115) for more information.*

Following Page:
The Dillingham Transportation Building, at 735 Bishop Street, is a productive office building as well as a historic attraction in the heart of downtown Honolulu. In a Mediterranean style, with a decorative facade featuring transportation themes in bas relief, the building memorializes one of Hawai'i's foremost transportation moguls—Benjamin Franklin Dillingham.

DILLINGHAM TRANSPORTATION BUILDING

'Ewa Plantation

James Campbell (1826–1900) was born in Londonderry, Ireland, and ran away to sea at age thirteen to live the life of a reckless adventurer. After many years of travel he eventually found his way to Maui, where he started a sugar plantation in 1861. Campbell's entrepreneurial spirit then led him to O'ahu and the plains of 'Ewa, where he envisioned a ranch with the finest livestock in the islands. In 1877, Campbell purchased 41,000 acres at Honouliuli, west of Pearl Harbor, for $95,000. The dry, barren plain called "Campbell's Folly" was so named because it could scarcely support five head of cattle. While his contemporaries scoffed, Campbell pursued a plan to irrigate the land and make it productive for farming and cattle raising. An experienced California well driller, James Ashley, came to Hawai'i to survey the situation. Ashley drilled his first well to a depth of 273 feet on Campbell's 'Ewa property in September 1879. The well was named Wai-Aniani, meaning "Crystal Waters," by the Hawaiians. It brought new life to the 'Ewa plains and flowed steadily for sixty years, until it was sealed by the City and County of Honolulu in 1939. The site of this first artesian well is marked by a plaque on a lava rock situated some 150 feet from the original well site on Fort Weaver Road.

Anticipating a burgeoning sugar industry in the islands, Campbell put his energy into creating a highly profitable, model sugar cane milling plant at 'Ewa. The entire area, including the plantation homes, became a center of industry. Two-lane roads wove an intricate pattern through an endless sea of sugar cane to reach the plantation villages of Fernandez, Renton, Varona, and Tenney. The villages were self-sufficient, with shops, schools, churches, medical facilities, plantation offices, a sugar mill, post office, restaurants, saimin stands, and even a graveyard. During harvesting, truckers worked twenty-four hours a day. The villages provided their own entertainment in the form of carnivals, musical events, and sports. Their only real link to the rest of O'ahu was the railroad system.

Today, 'Ewa Plantation is one of the important historic sites in Hawai'i yet to be afforded the privileges and protection of state and national historic landmarks. A new vision for the area, that of a second city of Kapolei, is now in its early stages of development. 'Ewa Plantation, with the support of the Friends for 'Ewa and the Historic Hawai'i Foundation, is struggling to preserve its historic identity in the face of an encroaching wave of development. The Hawaiian Railway Society is working to restore portions of the track, flatcars, diesel engines, and locomotives and to make them operational so people can once again experience train transportation from the old plantation days. The future of one of the last vestiges of a golden era of plantation life in Hawai'i remains to be seen.

• *See Historical Notes (page 115) for more information.*

EWA PLANTATION

'Ewa town, approximately 25 miles west of Honolulu, is the last remaining sugar plantation village on O'ahu. The 'Ewa villages along Renton Road, off Fort Weaver Road, were once a thriving plantation community during the height of O'ahu's sugar industry. Efforts are currently under way to preserve the site as a National Historic Landmark.

Falls of Clyde

The *Falls of Clyde*, built in 1878 by the famous River Clyde shipbuilding firm of Glasgow, Scotland, was one of nine ships named after waterfalls in Scotland. She is the last four-masted, iron-hulled, full-rigged sailing ship and the only sailing oil tanker afloat today. Built to the British medium clipper model—"deep watermen," as they were called—she carried cargo between Britain and India and also journeyed to Australia, New Zealand, Bangkok, Hong Kong, Shanghai, Oregon, California, and New York. Similar to other British vessels of the period, she was well built—with a combination of speed, cargo capacity, and ease and economy of operation. Now over one hundred years old, the *Falls of Clyde* led an interesting series of lives.

In her original career she served markets around the world as a tramp freighter, carrying lumber, jute, cement, wheat, and other bulk cargo. In 1898 she became the ninth ship of the original Matson fleet and spent her midage in Hawai'i's trans-Pacific passenger and freight trade business. William Matson, a merchant seaman, spent about $15,000 converting the freighter into a passenger ship equipped with deckhouse, charthouse, living quarters, and a small wooden shelterhouse on the poopdeck. Her rigging was changed so that a smaller crew could manage her. From 1899 to 1907, the *Falls of Clyde* was a cargo and passenger vessel between Hilo and San Francisco. She made sixty voyages, with an average sailing time of seventeen days.

When the Matson Navigation Company was formed, the *Falls of Clyde* became one of its seven

The *Falls of Clyde*, in Honolulu Harbor at Pier 7 next to the Hawai'i Maritime Center, is the only remaining four-masted, full-rigged sailing ship in the world. The ship and maritime museum are open daily, 9:30 AM–4:30 PM.

original ships. She was converted into a sailing oil tanker in 1907, when Matson became a pioneer in the sea transportation of oil. Ten large tanks were built into her hull, and heavy-duty pumps and a second steam boiler to operate them were installed. Matson then sold her to Associate Oil Company, and she sailed mainly between Gaviota, California, and Honolulu carrying 756,000 gallons of oil on each westward trip. At Honolulu she was loaded with bulk molasses for the return trip to California. She remained an oil tanker until 1921 and made voyages to Texas, Denmark, Argentina, and Panama.

In 1922 in San Pedro, California, the *Falls of Clyde* was converted to a barge. Her lower masts were cut down and the bowsprit removed. She was then towed to Ketchikan, Alaska, where General Petroleum Company used her as a floating fuel barge. Several company managers lived on board with their families at different times until 1958. A private owner, William Mitchell, purchased her in 1958 and offered her to various groups for sale as a historic vessel. Several cities tried to purchase her for museum purposes but were unsuccessful. Left with no alternative, the owner was about to sell her to Vancouver, British Columbia, where she would be scuttled as a breakwater. In 1963, however, a group aided by the Matson Navigation Company raised over $25,000 to buy the *Falls of Clyde*. She was towed to Honolulu by the U.S. Navy tug *Moctobi*, and the task of restoration was begun. Community support for the project was enthusiastic, and most of the work was accomplished by volunteers. Over the years the ship was restored to her original 1878 state as a full-rigged sailing vessel. This maritime relic was first operated at Pier 5 by the Bernice Pauahi Bishop Museum and later turned over to the Hawai'i Maritime Center. The ship was moved to Pier 7, where she lives out her final life as a floating museum and a rare historical treasure enjoyed by visitors from around the world. Alongside the pier is the newly built Pacific Maritime Center, which houses a museum, gift shop, and information center.

Foster Botanic Garden

Foster Botanic Garden origins go back to the 1850s, when a young medical doctor from Germany landed in Hawai'i in search of a climate beneficial to his poor health. Dr. William Hillebrand was born in Nieheim, Westphalia, Prussia, on November 13, 1821. He received his medical degree in Berlin, where an important part of his instruction was in medicinal botany. He began his medical practice in Prussia but fell ill, supposedly from tuberculosis, and was advised to seek a warmer climate.

Hillebrand's search led him to Australia and then to the Philippines, where he again fell ill. He moved on to Hawai'i, arriving in 1851, and found refuge in the home of Dr. Wesley Newcomb. The climate appealed to Hillebrand, but more so did the loving care he received from Newcomb's stepdaughter, Anna Post. Hillebrand recovered quickly and on November 16, 1852, he and Anna were married.

Hillebrand joined the Royal Hawaiian Agricultural Society in 1853. One of the society's objectives was to establish a botanical garden and nursery in Honolulu. Hillebrand became a strong advocate of the plan, and thus began a lifelong devotion to horticulture and the introduction of new plants and trees to Hawai'i. In 1855 he leased land in Waikahalulu from Queen Kalama, through her agent and uncle, Chief Charles Kana'ina. Because this acreage was not on the main road, Hillebrand also purchased an 11-foot right of way along the *makai* boundary, so that he could construct a carriage drive into his leasehold. It was on this land that Hillebrand began to assemble his botanical collection. He brought in new varieties of plants and made exchanges with botanical gardens in London and Berlin. He expounded on the need for an organized system such as a botanical garden, which could partake of the botanical collections of many other countries and export the unique island flora as well.

In 1857 the Royal Hawaiian Agricultural Society was granted 50 acres for a nursery in Nu'uanu Valley, near the country residence of John Young. The additional acreage for the botanical garden was donated by Mary Robinson Foster, a neighbor of Hillebrand, who bequeathed her home and garden to the City of Honolulu.

Mary Foster had engaged the services of Dr. Harold L. Lyon to bring some order to her garden, which had been neglected while she lived with her sister after the death of her husband. Lyon had come from Hastings, Minnesota, to work as an assistant pathologist on diseases of sugar cane. He became head of the Department of Botany and Forestry and director of the Hawai'i Sugar Planters' Association's (HSPA) Manoa Arboretum (now Lyon Arboretum). He agreed to assist Foster with her garden and persuaded her to lease part of her land to HSPA for an experimental station and nursery. Lyon brought along a friend and colleague, Joseph F. Rock, a botanist at the College of Hawai'i, to help identify and classify the many species of plants in the garden. By 1920 they had restored the garden to its former splendor and succeeded in identifying a total of 145 different species. Before Foster died, she arranged for transfer of her property to the City of Honolulu, including $10,000 for improvements. Her only stipulation was that it be called "Foster Park." She told Lyon that she had seen her husband, Captain Foster, riding his favorite horse at night in the garden. If it was reserved as a park, she believed, she and her husband would always have a place to return to.

After acquiring the property, the city turned its management over to HSPA and agreed to contribute $2,000 per year for its development as a botanical garden. Lyon was appointed director, and he used the $10,000 bequeathed by Foster for the construction of a glasshouse in which to begin the garden's orchid collection.

Foster Garden opened on November 30, 1931. It grew and expanded over the years, taking on additional acreage from private donations and from the City of Honolulu. Part of the land was sacrificed for the construction of the Lunalilo Freeway in 1961. The Friends of Foster Garden, a private, nonprofit organization, was formed in 1963 to assist in the preservation and public awareness of the gardens. The Outdoor Circle promotes the garden's educational programs and receives visitors numbering over 100,000 each year. The Harold L. Lyon Garden, featuring many of the orchids Lyon had propagated in a naturalistic setting, opened in 1964. The Prehistoric Glen was established and dedicated a year later.

Foster Botanic Garden today consists of three sections: the original 5 acres given to the City of Honolulu by Foster, which is known as Foster Park; 8.5 acres purchased by the city called the Garden; and the separate 7-acre Queen Lili'uokalani Garden, located *mauka* of Foster Garden along Nu'uanu Stream, which was added in 1960. Collectively, the park and two gardens encompass over 20 acres.

Foster Botanic Garden, with its unsurpassed collection of native and introduced plants, is a mecca for the student of botany. It is a place of beauty and serenity for the casual visitor strolling beneath the shade of such exceptional trees as the baobab, cannonball, kauri, and sausage. Throughout the garden one finds labeled varieties of palm, heliconia, ginger, and the other fascinating plants of the nation's largest collection of tropical flora. Foster Botanic Garden preserves a unique part of Hawaiian history and culture not found in other historic landmarks.

• *See Historical Notes (page 116) for more information.*

Thousands of rare and interesting plants inhabit this garden near downtown Honolulu, at 180 North Vineyard Boulevard. An impressive collection of orchids, trees, and tropical foliage makes Foster Garden a leading botanical site in the nation. Open daily, 9:00 AM –4:00PM. Guided tours are available.

Hawaiian Electric Company Building

The Hawaiian Electric Company is one of the few existing companies that can trace its beginnings to the era of the Hawaiian monarchy. The company formed a partnership in May 1891 and incorporated October 13, 1891, eight months into the reign of Queen Lili'uokalani, Hawai'i's ruling monarch. The company began operations in a one-story brick building where the Pacific Trade Center now stands. It originally offered ice and cold-storage services, which continued until home electrical refrigeration became common after World War II.

The present Hawaiian Electric Company building on the corner of Richards and King streets was built in 1927 at a cost of $750,000. Reports of the project boasted that its founders "spared neither expense nor effort to provide a structure adapted not only to its own need, but to the locality in which it was situated."

The architects credited with this handsome building were York and Sawyer of New York, who worked with local architects from the company of Walter L. Emory and Marshal H. Webb. Its cream and buff exterior blends harmoniously with the Old Federal Building (see U.S. Post Office, Custom House and Court House), which had also been designed by York and Sawyer some years earlier.

The Hawaiian Electric Company building was done in a Spanish Colonial style, incorporating reinforced concrete, steel floors, doors of metal, and a tinted California stucco exterior. Until the mid-1940s the first floor was used to display cases for the latest electrical appliances. The upper three floors were occupied by engineering and administrative offices. The remaining space was rented to independent business professionals.

The building is a fine example of the blending of elegance and function. Some unique features include a three-bay, groin-vaulted portico at the Richards Street entrance, which was handpainted with mythic figures by artist Julian Jarnsey of San Francisco. A matched ceiling once adorned the entrance between King and Merchant streets but was painted over for reasons unknown.

The first-floor ceiling was richly detailed with hand-plastered and painted moldings by artist J. Rosenstein. Its arched windows with Churrigueresque column supports (a style introduced into Spanish architecture by José Churriguera) are one of the unique aspects of the building. Another interesting design feature on the exterior is the corner cupola on the Diamond Head side, with its polygonal tiled roof and lantern.

• *See Historical Notes (page 117) for more information.*

The impressive and uniquely designed Hawaiian Electric Company building, located at 900 Richards Street between King and Merchant streets, is a prominent site in Honolulu's most historic district. The artistry of its interior and exterior sets it apart from other commercial buildings of its day. Built in 1927, it was the second structure to be occupied by Hawai'i's first electric company.

Honolulu Academy of Arts

ANNA CHARLOTTE RICE COOKE, founder of the Honolulu Academy of Arts, wanted to establish a museum and an academy for arts that would be a "living, vital, and heartwarming place where people would be made aware of man's freedom of spirit" and where children could be educated about art. After looking for an adequate building in which to house their art collection, and finding none, she and her husband donated their homesite on Beretania Street and made plans to construct a new building. They commissioned New York architect Bertram Goodhue to design the building. Goodhue chose a combination of Chinese and Spanish architectural styles, with galleries grouped around open courts in a setting enhanced by surrounding mountains, trees, and shrubs. Though Goodhue's plans were well received, he died shortly after submitting them and the project was continued by a colleague, Hardie Phillip, of the Goodhue firm. Phillip introduced modifications according to the Cookes' request to make the structure more Hawaiian. He borrowed the peaked roof from the Polynesian house and the informal lanai from what he took to be an early missionary adaptation of the New England veranda.

In December 1926 the building was completed, and on April 8, 1927, it opened with a simple and inconspicuous celebration. The Royal Hawaiian Band played music and Hawaiians entertained with songs. To Anna's astonishment, fifteen thousand people visited the museum over the remaining three weeks of April.

In 1929, *Paradise of the Pacific* wrote this of the new Academy: "It grows from the Hawaiian soil as the cocopalms, the ti and banana which throw shadows upon sloping roof and whitewashed walls." Visitors marveled at the lovely museum and grounds, as well as at the art inside. The Chinese Court held a natural flagstone pool, surrounded by a graceful building with a sloping tiled roof and broad lanais covered by an extended roof, providing deep eaves for shade. The Spanish Court featured a tile pool and natural flagstone flooring. The Hawaiian Room displayed relics of royalty and the chiefs. There was even a music room and a hall devoted to modern art.

Today the Honolulu Academy of Arts is noted for the quality and diversity of its collection of Oriental, American, and European art. It has one of the finest Asian collections in the United States. Exhibits in its thirty galleries are beautifully enhanced in a setting of sparkling pools in open courts shaded by trees and foliage. The compatible blend of art, architecture, and natural surroundings creates a peaceful, dignified atmosphere throughout the halls and galleries. In expanding, the Academy introduced art education, films, lectures, special exhibitions, and other cultural activities. Most recently, it acquired the nearby Linekona School as the site of its art education program.

The Honolulu Academy of Arts is a tribute to a remarkable woman whose dedication to the cultural and educational experience of her island home has enriched Hawai'i and the world. In a 1926 brochure, Anna Rice Cooke's hopes for the Academy were expressed in these words:

"That our children of many nationalities and races, born far from the centers of Art, may receive an intimation of their own cultural legacy and wake to the ideals embodied in the arts of their neighbors. . . . That Hawaiians, Americans, Chinese, Japanese, Koreans, Filipinos, North Europeans, South Europeans and all the other peoples living here, contacting through the channel of Art those deep intuitions common to all, may perceive a foundation on which a new culture, enriched by old strains, may be built in these islands."

• *See Historical Notes (page 117) for more information.*

Honolulu Academy of Arts *Jeannette Murray 1986*

This inner courtyard scene greets visitors to the Honolulu Academy of Arts, at 900 Beretania Street. Founded in 1927 by Anna Rice Cooke, the Academy features traveling exhibitions and a permanent collection of European, American, and Asian Art. For information on tours, call (808) 538-3693. Hours are Tuesday through Saturday, 10:00 AM–4:30 PM., and Sunday, 1:00–5:00 PM.

Honolulu Hale

On the corner of King and Punchbowl streets, directly across from Kawaiahaʻo Church, is a Mediterranean-style building that contrasts sharply with its New Englandish neighbors in the vicinity. Honolulu Hale, or City Hall, was built to house the entire city government. It opened its doors on November 26, 1929, and was officially named the Municipal Government Building, though it was never called by this name. The building was designed by a team of architects: Charles W. Dickey, Hart Wood, Robert G. Miller, Guy Rothwell, Marcus C. Lester, and John H. Kangeter.

Throughout the 1920s, local architects strove to incorporate traditional styles such as Italian, Spanish Mission, and Spanish Colonial into a unique style that would serve both functionally and climatically in the islands. Honolulu Hale is a perfect example of this historic phase of architectural blending, with Spanish Colonial and Italian architectural features highlighted by Hawaiian motifs.

On approaching the main entry on King Street, one sees three arched entryways below an open balcony that runs the length of the building. On the Punchbowl side, a large, six-story, medieval-looking tower with carved stone balconies almost appears to be a separate structure. From this side one sees an entirely different facade with balconies, loggias, deeply recessed windows, open and closed arched windows, cast-iron grillwork, and double-hung windows. Entering from King Street, one finds an open courtyard with a grand staircase leading to the fourth-floor mezzanine. Part of the central courtyard is open to the sky; the rest is covered by an ornately coffered ceiling with frescoes that are among the most intricate and noteworthy features of the building. They were designed and painted by artist Einar Peterson of California, who utilized American Indian and Moorish design elements. Individual drawings of birds, fish, canoes, and other depictions can be found along the main beams. Carved stonework is found on the columns, interior and exterior balconies, and decorative trim throughout the building. This work is credited to Italian sculptor Mario Valdastri, who had been brought to Hawaiʻi by Julia Morgan to work on her YWCA project. Valdastri stayed in Hawaiʻi for forty years, then retired to Italy. He later returned to Hawaiʻi to work on the restoration of ʻIolani Palace.

Around the perimeter of the courtyard are interior, government offices. With the growth of the Honolulu government, Honolulu Hale has retained its distinction as a government headquarters that continues to house the mayor, city council, and principal staff agencies. But Honolulu Hale is as much a social and cultural center for the people of Hawaiʻi as it is a government building. Its bright, airy courtyard interior is host to many public exhibits and year-round community activities. Each Christmas, the front grounds are transformed into a wonderland of colorful holiday figures, decorations, and lights. Inside is an exhibit of Christmas trees and wreaths. Nowhere in the world will you find the likes of its huge Hawaiian Santa Claus, with boots off and rolled-up pant legs, soaking his feet in the cool waters of a fountain and hailing passers-by with the *shaka* sign, a local gesture of greeting.

Honolulu's City Hall is a delightfully decorative building located on the corner of King and Punchbowl streets, across from Kawaiaha'o Church. Its ceiling frescoes are among its many outstanding features. It continues to serve government purposes while providing a site for artistic and cultural events. Honolulu Hale is open to the public Monday through Friday.

'Iolani Palace

'Iolani Palace
Honolulu, Hawaii

KING KALAKAUA (1874–1891), THE Merrie Monarch of the Pacific, was as flamboyant and affable a socialite as he was a shrewd politician. Beloved by the Hawaiian people for his warmth, gaiety, and musical talent, he was equally despised by a cadre of non-Hawaiians who were anxious to see an end to the "absurd monarchy." It was inconceivable to Kalakaua—a lover of grandeur, luxury, music, and dance—that the old palace, built by Governor Kekuanao'a for his daughter, Princess Victoria Kamamalu, could serve as his royal residence. It was old, shabby, and termite-ridden. To Kalakaua, it represented the growing impotence and insignificance of the monarchy.

King Kalakaua's reign brought a new era to the island kingdom. Inspired in his travels by the affluence of his royal contemporaries, he went ahead with plans to build his royal palace. Its design has been described as "American Florentine" and "American Composite" (Walter F. Judd, *Palaces and Forts of the Hawaiian Kingdom*). The cornerstone was laid in 1879 during an elaborate Masonic ceremony. The building consists of two stories, with a basement, six towers, and lanais or porticos on all four sides, with Corinthian columns. The ornate interior displays elaborate ornamental plaster cornices and decorative ceilings. From a large central hall, an impressive staircase of koa leads to the upper story. Throughout the interior the finest of rare Hawaiian woods—koa, kamani, kou, and ohia—as well as Douglas fir, American walnut, and Oregon white cedar are used. It was originally

---

'Iolani Palace, located on South King Street between Richards and Punchbowl streets, is perhaps the most impressive site in historic Honolulu and a favorite attraction for visitors and residents. Tours are available by reservation Wednesday through Saturday, 9:00 AM–2:15 PM, every fifteen minutes. Call (808) 522-0832. The grounds are open to the public every day of the week.

'IOLANI PALACE

furnished with many European and American effects, some gathered during the king's travels or given to him by European leaders. Portraits of former kings, queens, and heads-of-state lined the walls. The southeast corner of the main floor served as an elegant throne room, where gala evenings of dancing to the music of the Royal Hawaiian Band were held. An accomplished musician himself, the king composed "Hawai'i Ponoi," which replaced an adaptation of Britain's "God Save the King" as Hawai'i's national anthem.

The palace's exterior was of plastered brick and iron, with concrete block trimmings painted in a light sand-colored hue. It took four years, three consecutive architects, and over $350,000 to complete. The exquisite crystal chandeliers, first illuminated by gas, were converted to electricity in 1887, even before the White House in Washington, D.C., could boast of this modern innovation. One of the earliest telephones installed on O'ahu was a private wire from the palace to the boathouse in which the royal yacht was kept.

From 1895 to 1968, 'Iolani Palace was the seat of the government for the Republic, the Territory, and the State of Hawai'i. On its front steps the formal ceremony of transfer of the Hawaiian Islands to the United States was enacted and the Hawaiian flag, emblem of an independent nation, was lowered and replaced by the American flag.

'Iolani Palace was vacated in 1969, when the new State Capitol was completed. It was in a dreadful state of deterioration after years of abuse and neglect. In 1970 plans for restoration were developed and funded by the state. Work began under the supervision of a group known as Friends of 'Iolani Palace. The project involved replacement of termite-ravaged structures, restoration of the entire palace to its original decor and design, and recovery of numerous artifacts and furnishings. Articles that could not be recovered were meticulously replicated. The most significant of the pieces returned were the thrones and golden crowns of King Kalakaua and Queen Kapi'olani, which had been held by the Bishop Museum.

Today the palace stands distinguished and resplendent after a commendable feat of restoration. So scrupulous are the efforts of the Friends of 'Iolani Palace to preserve it that touring visitors are required to wear cloth booties over their shoes in order to protect the floors. Thanks to the efforts of the Friends, visitors can experience the grandeur of Hawai'i's royal palace and the legacy of the Hawaiian monarchy.

• *See Historical Notes (page 118) for more information.*

ʻIolani Barracks

ʻIOLANI BARRACKS, ORIGINALLY CALLED Hale Koa, "Soldier's House," was designed by Theodore C. Heuck in 1866 as a fort to accommodate as many as 125 *koa*, soldiers of the king. Heuck, who is usually considered to be Hawaiʻi's first architect, had among his other accomplishments the first Queen's Hospital building and the Royal Mausoleum in Nuʻuanu Valley. Hale Koa was completed in 1871, ten years before ʻIolani Palace was built. It originally stood on a site now occupied by the new State Capitol behind the palace, across Hotel Street, which was formerly known as Palace Walk. It was moved from there in 1965 to its present site on the grounds of the palace to make room for the Capitol.

The diminutive, castle-like structure is a coral block building with 18-inch thick walls, high, barred windows, and a slate roof. Coral blocks were cut from reefs and some were obtained from the old printing house at the Mission Houses. Rumor had it that a secret tunnel connected the barracks to the palace, but this tunnel has never been located. The barracks was built to shelter Kamehameha V's household troops, whose primary responsibilities included guarding the king and the treasury, firing cannon salutes, and appearing in royal parades and ceremonies.

When the monarchy was overthrown on January 17, 1893, the King's Guard, then entitled the Queen's Guard under Queen Liliʻuokalani, surrendered to the Citizen Guard Forces and was disbanded once and for all by the provisional government. The barracks was then used for munitions storage. The Citizen Guard took over the barracks during the provisional government period (1893–1894), and the National Guard of Hawaiʻi headquartered there during the life of the Republic of Hawaiʻi. The territorial government took it over in 1899 and used it for office and storage space. After annexation of Hawaiʻi by the United States, ʻIolani Barracks became the possession of the U.S. War Department and was used by the Quartermaster Corps until 1917. After renovations in 1920, it became a service club for about a decade. In 1929, after another renovation, various government offices occupied it until 1943, when plans were announced for a military museum. The museum proposal never materialized and the building was renovated again for office use. In 1931 it was returned to the Territory of Hawaiʻi. It was used throughout World War II as the Hawaiʻi National Guard state headquarters. When the Guard moved to Fort Ruger in 1950, the barracks was utilized by territorial and state government agencies. In 1961 it was turned into a state storage area.

With the development of the State Capitol complex in the 1960s, the barracks was condemned and, in 1962, abandoned. In 1964–1965 the building was demolished and the coral shell moved and reconstructed, stone by numbered stone, in the northwest corner of the palace yard. Large sections of its walls were broken down and the original coral blocks were reset.

• *See Historical Notes (page 119) for more information.*

Iolani Barracks
Honolulu, Hawaii

Next to 'Iolani Palace stands the fortresslike 'Iolani Barracks, which was once occupied by the royal household guard. The present structure is actually a reconstruction of the original barracks, which was located behind the palace on the present site of the State Capitol. Today, the building contains the Palace Shop and a ticket office.

ʻIolani Palace's Coronation Pavilion

What is more commonly referred to today as the "Bandstand" on the ʻIolani Palace grounds was the coronation pavilion built in 1883 for the crowning of King Kalakaua and Queen Kapiʻolani. The king had traveled extensively throughout the world and was influenced by the royal customs and protocol of other countries. Upon returning from his travels, he was determined to be properly crowned as Hawaiʻi's king, even though he had held the title since 1874.

The coronation pavilion was originally built at the foot of the main staircase of ʻIolani Palace, surrounded by an amphitheater for the inaugural guests and connected to the palace by a platform. It was known as Keliʻiponi Hale and was constructed in an octagonal shape by Messrs. Buchman and Rupprecht, two artists who had recently arrived in the islands. Observers described it as the "finest specimen of this kind of work that has ever been produced in Honolulu." Each of its eight sides bore the name of one of the kings of Hawaiʻi, from King Kamehameha I to King Kalakaua. The ceiling was decorated with fresco work and the Hawaiian coat of arms was painted in the center of the white network. Its eight upright columns, representing the eight islands of Hawaiʻi, supported the domed sheet metal roof. Eight panels on the outside depicted the coats of arms of different nations. A spike above the cupola suggested a primitive Hawaiian spear.

The pavilion was an outstanding achievement for its time and was representative of the popular Victorian trend that appeared in the palace and other official buildings. Coronation festivities lasted for two weeks and provided unsurpassed feasting, games, dancing, and celebrating for all participants. The festivities were sharply criticized by local conservatives, who considered them, in fact the whole coronation idea, a frivolous waste of time and money and a scandalous display of *hula* dancing and other heathen practices. Reports tell that the pavilion was removed to the southeastern corner of the palace grounds and the whole space from the palace veranda to the pavilion covered with canvas to provide the "most spacious and elegant ball-room ever seen in Honolulu." Unfortunately, rain began to fall and the canvas covering leaked, so the ball had to be moved to the throne room of the palace. Four days after the coronation, the *Advertiser* carried the story that the pavilion was to be utilized on the palace grounds as a bandstand.

On February 27, 1883, the pavilion provided special seating for Her Majesty Queen Kapiʻolani and other members of the royal family and ministers at a coronation luau. Finally, on April 20, the pavilion was moved to the west side of the palace, where it continues to serve as a bandstand. The Royal Hawaiian Band plays each Friday at noon to an audience of downtown workers, tourists, and passers-by. The pavilion also serves as the inauguration site for Hawaiʻi's governors.

The Coronation Pavilion
'Keli'iponi Hale'

Located in the left front quarter of 'Iolani Palace, near the Richards Street gate, this original pavilion was built for the coronation of King David Kalakaua. Today it is used as a bandstand by the Royal Hawaiian Band, which gives free concerts on Fridays at noon.

Kaka'ako Pumping Station

The uniquely designed Kaka'ako Pumping Station was constructed in 1900 for Honolulu's first professionally designed sewage disposal system. Turn-of-the-century Honolulu confronted a major problem with inadequate sewage management for a steadily growing urban population. Until this time, sewage disposal had been largely a matter of dumping raw sewage into roads and gutters, under buildings, or into backyards. Horse-drawn pump wagons were used to evacuate cesspools, but most other sewage seeped into the ground. These unsanitary conditions were of grave concern to citizens and city officials, who feared the spread of disease and a resurgence of bubonic plague, which had threatened the city in the late 1800s. In 1896 the Territorial Department of the Interior commissioned a well-known sanitary engineer, Rudolph Hering, of New York, to design the sewage disposal system. Hering had gained a reputation through his study of European sewage systems and was known as the "Father of Sanitary Engineering." He proposed a system whereby separate networks of conduits carried sewage and storm waters. This system is still used in Honolulu today.

Work on the sewage system project commenced in 1899 and led to the construction of the Ala Moana Pumping Station, later to be known as the Kaka'ako Pumping Station because of its proximity to that area. Three other pumping stations were subsequently built in the same area. The station was designed by O. G. Traphagen, an architect from Duluth, Minnesota, who had also designed the Moana Hotel, the Judd Block, and the Hackfield Building. Traphagen's meticulous attention to detail and emphasis on artistic, yet functional, design features are apparent in this handsome public works building. Its Industrial Romanesque style displays rusticated walls of cut blue-stone, a triple-arched semicircular window, and an octagonal hip roof made of concrete, with a projecting gable. Behind the pumping station is a 76-foot smokestack made of scored cement over brick.

The Kaka'ako Pumping Station was used to house steam-powered pumps that carried sewage through a force main out to sea some 1,200 feet from shore. A small addition to the building was constructed in 1925, and in 1940 a new pumping station was built on the southwestern side of the original building. The 1900 structure was closed down and then converted into a facility for machine repair and maintenance. Another station was recommended just four years later when the 1940 station had already proved inadequate for the tremendous population increase in Honolulu after World War II. A third, larger station was completed in 1955, and a fourth was added in 1980.

Today the Kaka'ako Pumping Station lies empty and unused as it awaits restoration and transformation as one of the preserved buildings on a campus dedicated to historic preservation and education. The Kaka'ako Heritage Education Center will feature a public campus to serve as the center for historic preservation in Hawai'i. The project will include a community multipurpose room, an information kiosk, classrooms, libraries, administrative offices, and exhibit rooms. Completion is expected sometime in 1994.

• *See Historical Notes (page 119) for more information.*

Following Page:
Located on Ala Moana Boulevard and Keawe Street, this was the main pumphouse for Honolulu's first sewage disposal facility. Plans are under way to develop the site for the Kaka'ako Heritage Education Center, which will establish a permanent address for historic preservation in the State of Hawai'i.

Kamehameha V Post Office

As the size of Hawai'i's population grew, the need for a more organized, government-regulated postal service became pressing. The *Polynesian* in August 1859 reported that stamps had become a popular, even fashionable, incentive to writing, as well as a token of respectability. The earliest postal service in Hawai'i was authorized in 1850 and was actually a business concession operated by H. M. Whitney, who served as Hawai'i's first postmaster until 1856. The first post office was located in the original Honolulu Hale, not to be confused with the present day Honolulu Hale, or City Hall. In 1868 funds were appropriated for a new building, although rumors circulated that the legislature really wanted a new office for the government press. The original plans, which called for a two-story structure built from the coral stones of the old press building, proved infeasible and a new construction material was introduced—concrete. J. G. Osborne, a skilled brickmaker from Yorkshire, England, noted for his concrete block construction, brickmaking, and the use of corrugated iron roofing, handled the project.

The new post office was the first structure to be built entirely of precast concrete blocks with iron reinforced bars. Many doubted the suitability of the materials, and progress lagged as the concrete blocks had to be constantly wetted down to make them dry slowly in order to produce a firm and solid structure. The new building opened for business on March 21, 1871, and on April 25 the *Hawaiian Gazette* took over the upper floor and portions of the ground floor, confirming suspicions about the real intent to use the building as a government press office.

The post office was named the Kamehameha V Post Office to pay homage to the king responsible for its construction. Progressive for its time, the construction included sidewalks running from the front of the building down to the Advertiser building. It was considered "a model of neatness, regularity, and dispatch." Its large veranda at the front provided shelter from the rain and a place for posting notices and advertisements. To increase efficiency and expedite delivery, a ladies' window, a Hawaiian window, an all foreigners' window, a Japanese window, a Portuguese window, and a stamp window were included. Not until 1894 did the post office take over the entire building for postal use.

In 1900 an addition was built along the back of the building on Bethel Street, and on June 14 of that year the post office became a unit of the U. S. Postal System. This led to a deluge of mail from people wanting last-day cancellations by the Republic of Hawai'i and first-day cancellations by the United States.

By 1922 the post office had outgrown its headquarters and was moved to the Federal Building on the corner of King and Richards, known today as the U.S. Post Office, Custom House and Court House building. The old building became a postal substation, with the territorial tax office occupying the remainder of the space until it moved to a new building on Queen Street in 1939.

Next to the Kamehameha V Post Office is a small park, a project of the Honolulu Garden Club and Outdoor Circle. It is a memorial to Alan S. Davis, whose significant contributions helped shape downtown Honolulu. The fountain in front, surrounded by Caribbean royal palms, is made of old granite from China which came to Hawai'i as ballast on ships.

The historic Post Office is currently undergoing transformation into a 120-seat theater and permanent home for the Kumu Kahua Theater Group. It will also serve as the new headquarters for the State Foundation on Culture and the Arts.

On the corner of Merchant and Bethel streets in downtown Honolulu stands the first post office building in Hawai'i, built in 1870–1871. Now a restored building on the National Register of Historic Places, it is currently being prepared for use as a theater.

• *See Historical Notes (page 119) for more information.*

THE KAMEHAMEHA SCHOOLS

PRINCESS BERNICE PAUAHI BISHOP, great granddaughter of King Kamehameha I, inherited lands by virtue of her lineage and through the wills of various descendants of the king. Her largest inheritance came from her cousin, Princess Ruth Keʻelikolani. At the time of Princess Pauahi's death in 1884, her estate amounted to over 400,000 acres, to which her husband, Charles Reed Bishop, added 60,000 acres of his own. In her will, Princess Pauahi left instructions for the establishment of the Kamehameha Schools—one for boys, to be built first, and one for girls. These schools were to be the sole beneficiary of her estate, bearing the name of her great grandfather, King Kamehameha I. After his wife's death at the age of fifty-three, Bishop devoted himself to the founding of the Kamehameha Schools, which, according to Princess Pauahi's will, was intended to provide "a good education with instruction in morals and such useful knowledge as would tend to make good and industrious men and women."

The first school site, in Palama—now the grounds of the Bishop Museum—was selected because it was outside the town proper and was level and accessible by King Street, a narrow dirt road that ran through the property. After nearly fifty years at its original site, the Kamehameha

The Girls School auditorium, designed by Charles W. Dickey, is one of the many beautiful structures on the campus of the Kamehameha Schools, located on Kapalama Heights in Kalihi, overlooking leeward Oʻahu. The Kamehameha Schools serves native Hawaiian children and is one of Hawaiʻi's leading private schools, with more than 3,100 students in kindergarten through grade 12.

Schools outgrew its quarters and the trustees voted unanimously to construct new facilities on the steep slopes of Kapalama Heights. The architectural firm of Bertram Goodhue Associates of New York was selected to work with local architect Charles W. Dickey. The result was a beautifully designed group of buildings for the School for Girls adapted to the sloping terrain, made even more impressive by the scenic views of the leeward coast of O'ahu. Built first on the highest ridge, the complex included three dormitories, a gymnasium, a hospital, classrooms, a dining hall, an auditorium, and a library. It opened September 13, 1931, and the buildings were lauded in *The Friend* as "some of the finest in Honolulu, the location possessing the advantages of cool climate, a surpassing view and freedom from the distractions of city life." The auditorium, constructed in 1935, is especially noted for its ornate masonry screens and group of three large murals in brightly colored floral patterns on the front facade.

The School for Boys complex, completed in 1941, consisted of six dormitories, an assembly hall, a dining room, an administration building (Bishop Hall), an infirmary, shop buildings, an auditorium, and faculty quarters. The Preparatory Department moved to Kapalama Heights in 1955.

Buildings continued to be added over the years, and today there are more than sixty-five in use, most bearing Hawaiian names. An impressive recent structure is the Bernice Pauahi Bishop Memorial Chapel, which opened in 1988. While its design reflects the simplicity of earlier Hawaiian churches, its elegant furnishings include a 3,200-pipe organ, built by J. Walker and Sons of Kent, England, twenty *kahili*, and a small koa settee that belonged to Princess Pauahi. On the upper portion of the exterior koa wood walls is the thirteenth article of Princess Pauahi's will, hand-chiseled in two-inch high letters: her instructions to the trustees to build and maintain her schools. Next to the chapel is the Heritage Center, also completed in 1988, which serves to preserve and display some of the personal belongings of Charles Reed and Bernice Pauahi Bishop. Opposite the chapel and the Heritage Center is a 40-foot memorial wall constructed with over two hundred lava stones preserved from the original Bishop Memorial Chapel. The surrounding area is beautifully landscaped with traditional Hawaiian plants and trees, including a young tamarind that was planted in 1986 as a tribute to Princess Pauahi, whose father planted a tamarind tree on the day she was born.

From modest beginnings, the Kamehameha Schools has grown into the second largest private school in the nation, with an enrollment of over 3,100 students. Each year over 40,000 additional individuals benefit from participation in nearly fifty educational programs conducted at the 600-acre campus and in communities throughout the state.

• *See Historical Notes (page 121) for more information.*

Kawaiaha'o Church

KAWAIAHA'O CHURCH, THE "MOTHER" Protestant church of Hawai'i, was preceded by four consecutive thatched churches that either burned or were replaced by larger structures. The idea of building a permanent house of worship for the Christian missionaries in Honolulu originated with Kalanimoku, the chief in rank next to Ka'ahumanu, the queen regent. After Kalanimoku's death, the quest to build a new church was taken up by Governor Boki, influenced strongly by Queen Ka'ahumanu. When she died in 1832, Kina'u, her successor, gave her full support to the church, as did King Kamehameha III, who had expressed a great desire for three things: a royal yacht, a royal palace, and a new church. Kamehameha III contributed $3,000 to the building of Kawaiaha'o Church, and the rest of the revenue came from other donations.

Reverend Bingham drew up the plans for the church, including in them a two-story structure with a cellar, galleries, pillars in front, and a bell tower. The district chiefs oversaw the work of a volunteer force of parishioners. Numerous pits were dug where coral was burned and ground into lime for mortar to secure the 14,000 coral blocks hand-hewn from the nearby reef. Using saws and axes, the native laborers and missionaries cut blocks of living coral weighing 1,000–1,200 pounds each, then hauled them on handcarts or oxcarts to the churchyard. Timber was floated from northern O'ahu to Kane'ohe, then carried by teams of men up the perilous trail of the Nu'uanu Pali, across which no wagon could travel. Five companies of men from the congregation worked in rotating shifts. An immense sandstone cornerstone quarried at Wai'anae and floated to Honolulu was provided by Chief Abner Paki.

The entire cost of the work, including labor, was about $30,000. Before the name Kawaiaha'o was adopted, the church was known as Great Stone Church, Honolulu First Church, the King's Chapel, and even the "Westminster Abbey of Hawai'i." Finally, in 1862, the name Kawaiaha'o became official. It was derived from a legend about Ha'o, an *ali'i* whose beauty was so *kapu* that only chiefs could look at her. Commoners who stole the privilege were instantly put to death. It is said that Ha'o came one day to bathe ceremonially in the pool, Ka Wai a Hao, "The Water of Ha'o," located near the site of the present church. The water was so shallow that a huge sacred stone had to be dragged into the pool to raise the water level so that Ha'o could completely immerse herself. This historic stone was moved to the grounds of the church, where it now rests in an artificial pool and is marked by a tablet as Ka Pohaku o Ka Wai a Ha'o, "The Stone of the Water of Ha'o." None of the previous churches on the site bore the name of this sacred pool.

Kawaiaha'o Church is still considered the center of worship for the Hawaiian people. Sunday services are now held in English and Hawaiian, with hymns sung and programs written in both languages. This great stone church, together with the three nearby mission houses and the old adobe schoolhouse, are mementos to the lifework of the missionaries in Hawai'i. Fortified by repeated restoration efforts, Kawaiaha'o Church continues to serve its congregation as actively as it did in the days when kings and queens and native parishioners worshipped together within its sturdy walls.

• *See Historical Notes (page 120) for more information.*

Following page:
This coral block structure, located at 957 Punchbowl Street on the corner of South King and Punchbowl streets near the Mission Houses, is the oldest Protestant mission church in Hawai'i still open for services.

Kawaiahao Church

KAWAIAHA'O CHURCH

La Pietra

On the lower slopes of Diamond Head, surrounded by trees and lush vegetation, stands a beautiful Mediterranean-style mansion called La Pietra—home to the Hawai'i School for Girls. It was constructed in 1921 for Walter F. Dillingham, the prosperous descendant of early missionary and seafaring families, and his wife, Louise. La Pietra was named after a 600-year-old Italian villa owned by Mrs. Dillingham's aunt and uncle, Mr. and Mrs. Arthur Acton, and was designed as a composite of several beautiful villas Louise had visited with her aunt while in Florence. Its facade recalls the de Medici villa in Florence. The two-story structure is built around a central court lined with arcades and set in a spacious garden. The Mediterranean influence is carried throughout in its plaster and stucco wall finish, red tile roof over wide projecting eaves with large rafters, and ceramic tile floors. Originally the interior of the house was simple and restrained, providing an appropriate setting for the family art treasures. A central courtyard with sandstone columns and a romantic water-lilied pond are some of its finest features. The stone for the columns was quarried at Barber's Point at low tide and turned on a lathe. The entry and central courtyard are constructed of Kahuku sandstone.

La Pietra was built in 1921 on land purchased from James Campbell, who had obtained it from King Lunalilo's estate in 1883. Louise Dillingham selected architect David Adler of Chicago to design the home in

This former home of Walter F. Dillingham, designed in the style of an Italian villa, is now the Hawai'i School for Girls at 2933 Poni Moi Road. It was once Honolulu's best-known private residence, where notable guests and celebrities gathered to enjoy the finest of Hawaiian hospitality. La Pietra also housed the Dillinghams' magnificent art collection.

1919. His reputation was already established in Chicago with his designs for Armours, McCormicks, and Marshall Field. Terraces had to be cut into Diamond Head, and basalt blocks were quarried at Moʻiliʻili and dragged up to the site. Adler directed the project from across the sea by telegram and did not see it until after it was completed in 1922. The main structure contained five public rooms around a central courtyard. It had two downstairs bedrooms and eight upstairs, along with a sitting room and a screened porch. There was also a kitchen, pantry, cloakroom, and vault. The walls were constructed of hand-cut stone, and four inches of plaster was added to create the pink stucco exterior. It cost an unbelievable sum of $400,000 to build.

The Dillinghams lived at La Pietra until their deaths in 1963 and 1964. They willed the property to Punahou School, but Punahou was unable to use it and decided to put it on the market for one million dollars. The problem was finding a buyer who could afford the expensive pricetag, which was actually far less than the house and land were worth. The building had fallen into disrepair and would require a great deal of work if it were to be adapted for use as a school facility.

The founder of the Hawaiʻi School for Girls, Mrs. Richard A. (Lorraine) Cooke, and Mrs. Garner Anthony were looking for a place to accommodate their growing school, which was six years old at the time. La Pietra seemed the perfect location, and through community support and generous donations the land and house were purchased. While the school continued to operate at Central Union Church, La Pietra was painstakingly restored and made ready for its new purpose. Renovation and adaptation plans were developed by architect John Tatom and his assistant, Tom Fanning, and the actual construction work was done by parent volunteers and Dillingham's Hawaiian Dredging Company. The renovation was finally completed in time for the 1969/70 school year.

La Pietra remains a monument to the Dillingham family and an excellent example of how preservation and modern adaptation are blended harmoniously on a historic site.

• *See Historical Notes (page 121) for more information.*

Linekona School

Linekona School traces its origins to 1895, when the Fort Street School divided into Kaʻiulani Elementary and Honolulu High School. The high school was renamed McKinley High School, and later the building was used for an elementary school when McKinley High School's campus opened on King Street. The elementary school was called Lincoln (Linekona) School and served as such until 1956. Linekona School gained recognition as the first site of the University of Hawaiʻi: the College of Hawaiʻi was located from 1908 to 1910 in the Maertens residence at the rear of the high school building. From 1957 to 1973, Linekona School was used as a school for children with learning difficulties and for classes in English as a second language.

Architecturally, the building was considered one of the finest of its kind in the islands. Imposing in size and design, it was composed of hollow concrete blocks shaped and colored to resemble Hawaiian blue-stone. It had eight classrooms, a library, an assembly hall, physics and chemistry labs, and a darkroom. Its designer, H. L. Kerr, was well known for other impressive structures: the McCandless Block, the Yokohama Specie Bank, and the Mission Memorial. He was noted for an eclectic style and an ability to imitate construction materials. The Linekona design is a splendid example of turn-of-the-century Georgian Revival style with Romanesque features. Linekona was built during a time of architectural experimentation when classic styles were chosen to denote permanence and stability, then uniquely adapted to island climate and atmosphere.

In the late 1970s, the aged and deteriorating Linekona School was abandoned and earmarked for demolition. Fortunately, an alternate plan was developed to save and restore the building to its original state. In 1986 the state Department of Land and Natural Resources awarded a 55-year lease of the site to the Honolulu Academy of Arts. A $2 million renovation and refurbishing project was carried out and Linekona reopened its doors in 1990 as the Academy's education center. Once again, the beautifully renovated Linekona School distinguished itself in the educational history of Hawaiʻi. Situated in a picturesque setting opposite Thomas Square, a short walking distance from the Academy of Arts, Linekona School perpetuates the artistic ambiance of the area.

• *See Historical Notes (page 121) for more information.*

Following page:
This two-story concrete block building located at 1111 Victoria Street, across from Thomas Square, was the first public high school in Hawaiʻi. It is now the education center for the Academy of Arts and features art classes and exhibits open to the public. It is the oldest public school building remaining in Honolulu.

LINEKONA SCHOOL

Lunalilo Mausoleum

"Lunalilo Ka Moi," Lunalilo the King, is inscribed on the face of a small chapel-like building nestled in a garden at the entrance to Kawaiahaʻo Church. This is the tomb of King William Charles Lunalilo, who stipulated in his will that he be buried in the Kawaiahaʻo churchyard and not in the Royal Mausoleum in Nuʻuanu Valley which housed the remains of the Kamehameha dynasty. A disagreement with this rival branch of the royal family had led to Lunalilo's decision. His mother, the High Chiefess Kekauluohi, was not included among the *aliʻi* whose remains were moved from the old mausoleum on the ʻIolani Palace grounds to the new mausoleum in Nuʻuanu. Outraged by this show of disrespect, Lunalilo vowed that he would be buried near his mother and his people in the Kawaiahaʻo churchyard. Aside from King Kamehameha I, whose resting place is unknown, King Lunalilo is the only Hawaiian monarch to be buried outside the Royal Mausoleum.

William Charles Kanaʻina, sixth king of the Hawaiian Islands who ruled under the title of Lunalilo, was the first elected Hawaiian monarch. When Kamehameha V died in 1872 without leaving an heir, the founder's dynasty ended. According to the terms of the constitution, the nation's legislators had to elect his successor. The election resulted in a victory for "Prince Bill" over his opponent David Kalakaua. A story is told that, as a boy, Lunalilo had always wanted to play the bass drum in the royal band, but, because it was not considered a proper thing for a high chief to do, his parents refused his request. After becoming king, Lunalilo's first act was to take the bass drum from the drummer and lead the royal band in a triumphal march around the palace grounds.

The mausoleum was designed by Robert Lishman, an Englishman who came to Hawaiʻi in 1871 from Australia to supervise the construction of Aliʻiolani Hale. The tomb is built of concrete with a slate roof in the form of a Greek cross, the same design used for the Royal Mausoleum. Its four gables are surmounted by crosses and in each, except for the front, are two narrow windows of stained glass. This was one of the first buildings to be constructed of concrete, the first being the Kamehameha V Post Office in 1870. The coffins of King Lunalilo and his father are encased in marble imported from Italy in a small room with red carpeting and three *kahili* standards.

• *See Historical Notes (page 121) for more information.*

This small gothic mausoleum built for King William Charles Lunalilo is located on the grounds in front of Kawaiaha'o Church at the corner of King and Punchbowl streets. It is one of the early concrete block buildings in Hawai'i, built in 1875. It contains the tombs of King Lunalilo and his father, Kana'ina.

McKinley High School

On busy South King Street, close to downtown Honolulu, stands the oldest remaining public high school still active in Hawai'i. Its spacious lawn is bordered by Chinese banyan trees, forming a central quadrangle around which stand the four original classroom buildings and the Marion MacCarrell Scott Auditorium, now the high school's administration building. An eight-foot bronze statue in the center of the oval drive represents the man who signed the resolution to annex Hawai'i as a territory in 1898, President William McKinley.

Originally named the Fort Street English Day School and begun by Rev. Maurice B. Beckwith in 1865, the public high school moved from the basement of the old Fort Street Church to the new Fort Street School on the corner of School and Fort streets in 1869. In 1895 it moved into the former palace of Princess Ruth on Queen Emma Street and was renamed Honolulu High School. As the high school outgrew the palace site, the board of education made plans for a new high school building on Victoria Street, then called Linekona School. In honor of the late president, who was assassinated in 1901, the school name was changed to McKinley High School. A statue of President McKinley was commissioned for $8,000. The 8-ton statue was shipped from New York to Honolulu and unveiled on February 23, 1911. By 1913 the building could no longer accommodate its growing enrollment, and the school moved to its present location on South King Street.

McKinley High School's buildings represent several fine examples of Spanish Colonial Revival architecture found in Hawai'i in the 1920s and 1930s. The first buildings were designed by Louis Davis and Ralph Fishbourne in 1922 and formally dedicated in 1928. Extensive use of terra cotta embellishment set them apart from such contemporaries as the Hawaiian Electric Company building and the YWCA.

As its curriculum expanded, the facilities increased from four buildings in 1923 to eleven in 1965. The first buildings erected on the new campus were the Commercial and Mathematics buildings, followed a year later by the Art and Home Economics buildings. The Marion MacCarrell Scott Auditorium, with a seating capacity of 1,114, was the largest theater in Hawai'i. It served both students and the community with performing arts and lectures. McKinley High School was the first school outside the United States to establish a chapter of the National Honor Society. The Fred Wright swimming pool, begun as a student project in 1923, enlisted the labor of more than a thousand students and the contributions of over five thousand alumni. The pool opened in 1938, with U.S. Olympic medalist Duke Kahanamoku of the class of 1910 hosting the ceremonies. The Senior Core Building, completed in 1940, was a WPA-funded project designed by Louis Davis and Vladimir Ossipoff. Numerous buildings were added after World War II.

From its humble beginnings in the basement of the old Fort Street Church, McKinley High School has developed into a model public high school and one of Hawai'i's finest educational institutions.

• *See Historical Notes (page 122) for more information.*

The Marion MacCarrell Scott Auditorium is one of several fine examples of Spanish Colonial Revival architecture on the campus of McKinley High School. Located at 1039 South King Street, it is the oldest public high school still in operation in Hawai'i and was the leading public school during the 1920s and 1930s.

Mission Houses

In the heart of Honolulu stands one of Oʻahu's most historic centers, where the life and work of Hawaiʻi's mission families is commemorated in several restored landmarks. Built by the Protestant missionaries who first settled in Hawaiʻi in 1820, the three mission houses and nearby adobe schoolhouse were the hub of the religious, political, social, and educational activities in the 1800s.

The missionaries' first homes were grass houses, "filthy, and infested with vermin," and the inhabitants were forced to sleep on the damp ground. Although the missionaries had an entire frame house built in New England, disassembled, and shipped to Hawaiʻi, King Kamehameha II would not permit them to put it up. He did not want foreigners building permanent structures, and the house these missionaries had in mind was larger and finer than his palace, a mere grass house. After much pleading, the king finally relented, adding, "When you go away, take everything with you."

The frame house was erected in 1821. It was the most impressive house in the islands and the source of disturbing rumors that circulated rapidly among the natives. The cellar, it was reported, was a place for storing firearms and gunpowder to be used for subduing the people and capturing the island. Kalanimoku, former treasurer and principal adviser to King Kamehameha I, built a house opposite the missionaries' with a bigger cellar to reassure his colleagues that the invaders would certainly meet with resistance. Another rumor held that a human sacrifice was buried at each of the four corners of the house, as in an ancient Hawaiian custom when a *heiau* or palace was built.

The missionaries built a printing house in 1823 which first produced an alphabet book and a Hawaiian hymnal. They printed primers, tracts, textbooks, reports, hymnals, and books of the Scriptures as fast as they could be translated. On May 10, 1839, the long-awaited Ka Palapala Hemolele, the Holy Bible, was printed in Hawaiian.

The Chamberlain House was constructed between 1828 and 1831. When it was completed, Levi Chamberlain and his family moved into three rooms on the ground floor. Chamberlain had come to Honolulu to serve as business agent for the mission, and his home came to be called the Depository, because here all the mission goods and property were received and stored. It was built of coral blocks cut from the nearby reef and was the largest of the three houses.

The three mission houses, headquarters of the Hawaiian Mission, were the scene of many significant events that shaped the future of the islands. Here the mission families taught school, preached the Gospel, and introduced a new culture and standard of living.

The Hawaiian Mission Children's Society, formed by descendants of the missionaries, recognized the historic significance of the buildings and took charge of their restoration and preservation. The Society continues to hold its annual meeting at the mission site. The entire complex, with the old adobe schoolhouse and cemetery, is a memorial to the early missionaries who lived, worked, raised their children, and buried their dead while laboring to convert and reeducate an entire nation. These brave pioneers traveled thousands of miles from home and relatives and all that was familiar to them to fulfill their goal of bringing "the worship of God to the Sandwich Islands."

• *See Historical Notes (page 122) for more information.*

CHAMBERLAIN HOUSE
Mission Houses Museum
Honolulu, Hawaii

The Mission Houses, located at 553 South King Street, were the first permanent structures built by the first mission company to come to Hawaii. The Chamberlain House, depicted here, was the third structure built and served as a home and storage house for the mission. Hours are Tuesday through Saturday, 9:00 AM–4:00 PM, and Sunday, 12:00–4:00 PM. The Mission Houses offers a historic walking tour of the area and a Living History program on Saturdays. For information, call (808) 531-0481.

Moana Hotel

The Moana Hotel "First Lady of Waikiki"

The Colonial-style Moana is Waikiki's oldest surviving hotel. It was the first high-rise building on Waikiki Road, now Kalakaua Avenue, and the first beachside hotel built on the Peacock premises, "where the surf was better than any other point" on Waikiki Beach. It was also the costliest and most elaborate hotel of its time. Several hotels preceded the Moana, including the Hotel Waikiki, the Waikiki Seaside, and the Sans Souci Hotel, which all dated from 1884. None of these earlier hotels are still standing.

From its beginning in 1901, the Moana catered to the "steamboat set" and was noted for its elegance and variety of beach sports, including surfboarding, outrigger canoe paddling, swimming, and sailing.

During the Moana's construction, Hawai'i was annexed and made a U.S. territory. The hotel was financed by the Matson Navigation and Territorial Company and was designed by O. G. Traphagen, an architect from Duluth, Minnesota, who also designed the Alexander Young Hotel, the Kaka'ako Pumping Station, the H. Hackfeld & Company building, and the first Archives building in downtown Honolulu. Traphagen was noted for his roof terrace design and for introducing mainland architecture to the islands.

The original Moana was four stories high and had 74 rooms. The interior was of oak and pine, and each floor was furnished in a different wood. The second floor was oak, the third floor was mahogany,

---

"The First Lady of Waikiki," as the Sheraton Moana Surfrider Hotel is nicknamed, is located at 2365 Kalakaua Avenue. Throughout the years this landmark hotel changed appearance many times. With the recent renovation completed in 1989, the elegance and charm of its 1918 Italian Renaissance style has been restored. The Moana continues to be a favorite vacation spot on Waikiki Beach.

and the fourth was maple. Fifth and sixth floors, added later, were done in koa and cherry. On the rooftop was an observation lanai. In 1918 two wings were added, designed in the popular Italian Renaissance style.

The Moana was innovative for its day, with an in-house electric plant to supply power and light, an elevator, and a laundry facility. A favorite feature was its beachside dressing rooms directly accessible to the surf, "so that in leaving the water a person was not obliged to walk along the beach in the cool air before changing" (*Hawaiian Annual,* 1901–1904).

In 1925 the Matson Navigation Company acquired the Moana and made changes to its Victorian design. The original Grand Salon, which had first served as a ladies' lounge, was converted to a bar and furnished with stained-glass ceiling panels removed from the Matson ship SS *Lurline*. A dining room and dance floor extended out over the ocean and saw many a night of ballroom dancing to Hawaiian melodies. A 300-foot pier once extended over the surf and was a favorite place for moonlit romantic strolls. On July 3, 1935, the famous radio program "Hawai'i Calls," originally broadcast from the Alexander Young Hotel downtown, moved to the Banyan Court of the Moana.

Over the years, the Moana endured several facelifts in keeping with popular trends of the times. From the original Victorian–Beaux Arts style it changed to Italian Renaissance, then to Art Deco and 1950s Modern. The most recent restoration began with a proposal in 1983 which, when completed in 1989 at a cost of over $50 million, brought back the 1918 style and merged the Moana with the Sheraton Surfrider Hotel. Restoration planner and principal architect Virginia Murison proposed that the building be totally rehabilitated rather than demolished and replaced by a new structure. This led to a search to discover the early appearance of the hotel from the few remaining plans and drawings, old photographs, markings on the floors where columns once stood, and documents in the basement so old they crumbled at the touch. The twenty-month-long restoration project featured the return of the exquisite Kalakaua Avenue porte cochere, street-front colonnades, a rotunda, the grand staircase, the Banyan Veranda, the Grand Salon, and the Roof Garden: all elements of the original hotel design were either restored or replicated.

A description of the Moana Hotel which appeared in the *Pacific Commercial Advertiser* in 1901 still holds true today: "Magnificent of exterior and interior, and bearing in every detail the stately outlines of the old Colonial period, the new hostelry rivals the finest hotels which are to be seen in most metropolitan cities on the Mainland or on the Continent."

• ***See Historical Notes (page 123) for more information.***

O'AHU RAILWAY AND LAND COMPANY

THE PRINCIPAL FORCE BEHIND THE CREation of the O'ahu Railway and Land Company was Benjamin F. Dillingham, who came to Hawai'i from Brewster, Massachusetts, in 1865 as a sailor and remained to become a successful businessman and developer. In the late 1870s, Dillingham became interested in a colonization project for the lands along the western and northern coasts of O'ahu. The area was dry and believed to be infertile and worthless. But, in 1879, James Campbell discovered that the lands could be irrigated with artesian well water and set about establishing his plantation (see 'Ewa Plantation).

Dillingham's first objective was to provide an efficient means of transportation for crops grown in 'Ewa and other parts of O'ahu. A railroad seemed to be the best solution, and fortunately the Hawaiian government shared his views on the importance of a rail system for the kingdom. In 1878 the legislature passed a law to promote the construction of railways, guaranteeing developers a profit of 5 percent per annum on the cost of their roads and equipment. The three major railway developers at the time were Dillingham on O'ahu, Capt. Thomas Hobron on Maui, and Samuel G. Wilder on Hawai'i. Dillingham's line proved to be the most successful of all.

In 1886 Dillingham went to England to secure capital, but he failed because British financiers were unsure of the stability of the Hawaiian kingdom. After returning to Hawai'i he obtained a 50-year lease on lands at both Honouliuli and Kahuku from James Campbell in 1889. Dillingham convinced the legislature to contract with him to build and operate a steam railroad for the carriage of passengers and freight. Although a previous attempt by Charles Wilson to construct a railway from Pearl River Lagoon to Niu had failed, Dillingham succeeded in obtaining a charter from the government of Hawai'i for the O'ahu Railway and Land Company in February 1889. Construction began immediately and he devoted himself to the project, acting as both financier and construction worker. Determined to keep a promise to his friends that on his forty-fifth birthday he would give them all a ride on his railroad, Dillingham brought in Hawai'i's first steam shovel to speed the work.

The first track of German-made steel rails was laid in August 1889, and by Dillingham's forty-fifth birthday a few miles of track from Iwilei Station along the Honolulu docks were ready for use. Some flatcars had arrived, but one essential item was missing: the locomotive had not yet arrived from Philadelphia. To fulfill his promise, Dillingham had to purchase outright from the Honolulu Department of Public Works a small saddle-tanker locomotive, supposedly used for pulling streetcars in Honolulu. He named it Kauila, meaning "lightning," and added "No. 6" for the age of his youngest daughter.

By November 1889 the tracks extended as far as Aiea, and the line was formally opened to the public on November 16—King Kalakaua's birthday. Four thousand people rode free that day in new passenger coaches pulled by the Baldwin locomotive, which had finally arrived.

Tracks continued to traverse westward around the island. They reached Pearl City by January 1, 1890, Wai'anae by July 4, 1898, and Kahuku, the farthest point, by January 1, 1899. Dillingham's vision of an agricultural boom in Hawai'i had become reality, and his plantation developments included the 'Ewa Plantation Company, Kahuku Plantation Company, O'ahu Sugar Company, and Waialua Agricultural Company. By 1895 sugar production on O'ahu reached a record of 21,000 tons per year, with nine plantations in operation. The OR&L carried almost all produce to Honolulu. The

The Hawaiian Railway Society, located at 91-1001 Renton Road in 'Ewa, preserves the last vestiges of this once prestigious company in a railway museum featuring several of the original locomotives and railway cars of the O'ahu Railway and Land Company. The museum offers train rides, a gift shop, and information about the early plantation and transportation history of O'ahu. For information, call (808) 681-5461.

railroad continued to prosper and was one of the few in the United States never to miss a dividend to its stockholders.

A private railroad car was built for the Dillingham family in 1890. In this car, Dillingham hosted former president William Taft on a ride to Haleiwa.

Pineapple followed sugar as a cash crop, and Dillingham built a branch track to Wahiawa, where James Dole had his pineapple plantations. In Honolulu, Dole was able to set up his cannery on land provided by Dillingham, and by 1914 the OR&L was carrying 32,000 tons of pineapple a year.

With the development of Pearl Harbor and Schofield Barracks came increased business for the railway. Passenger service steadily increased, with an all-time high of 2,642,516 passengers in 1943. Altogether there were 100 locomotives, 88 of which were in the plantation service. During World War II, the railway performed yeoman service by carrying supplies, munitions, troops, and defense workers. After the war, passenger and freight totals fell drastically with the increased numbers of motorized vehicles. On December 12, 1947, all railroad operations outside Honolulu were abandoned, and it was not long before operations in Honolulu ceased. For the sum of one dollar, the U.S. Navy took over operation of the track between West Loch and Lualualei Ammunition Depot at Nanakuli and ran the railroad until 1968. This section of the railroad was placed on both the State and National Registers of Historic Places in 1974.

• *See Historical Notes (page 123) for more information.*

Old Honolulu Police Station

The three-story wooden building with stucco exterior at the foot of Nuʻuanu Avenue replaced the first police station in Honolulu, Kalakaua Hale. By the 1930s the original brick building, which had been commissioned in 1886, was falling into disrepair after serving for years as a police station, jail, and even emergency hospital. Newspaper reporters at the time described its jail cells as something out of the Spanish Inquisition, and its facilities were inadequate for the growing city population.

In 1931 a new structure, designed by local architect Louis E. Davis, was built at a cost of $235,000 by contractor F. M. Dias. It was simply called the Honolulu Police Station, and the *Honolulu Police Journal* described it as "one of the most elaborate structures of its kind in America and certainly the most ornate building in the Territory."

In designing the station, Davis used a Mediterranean Revival style of architecture popular at the time. Its three stories contain a maze of hallways and small rooms in which Davis managed to include $40,000 worth of decorative tile, eleven tons of imported French roja alicante marble, Philippine mahogany doors, koa paneling, and a grand receiving counter of red marble in the lobby. The tilework in the lobby was considered to be some of the finest in the Hawaiian Islands.

The building's exterior consisted of terra cotta walls with tiles of red, blue, yellow, and black—giving it a lively, Mediterranean look that some early observers thought was more like a "gaily decorated hotel than an ordinary bastille."

In a $3.4 million renovation completed in 1987, the fifty-six-year-old building underwent interior and exterior restoration after having been abandoned by its two former tenants, the police department and the district courts. Its interior was a shambles. For years it had been used for the annual March of Dimes Halloween Haunted House, and there were coffins in the lobby and gloomy Halloween paint on the walls. The formidable task of restoring the classic structure was executed by architects Fred Sutter and Richard Abe, who tried to preserve as much of the old look as they could—with none of the original blueprints to work from. Some segments of black marble had been replaced with wood painted to look like marble. The ceiling murals had to be repainted by a mural artist and the ceramic tile was restored by a ceramic artist. The old signs above the courtroom, an iron gate from what was once a holding cell on the third floor, and all of the fresco work on the ceilings and some of the walls have been retained. After restoration the building was occupied by several city and county offices, with the property management branch on the first floor, the rehabilitation loan branch in the basement, and the Finance Department's real property assessment staff on the second and third floors.

The building's foyer has a 23-foot-high ceiling. On the exterior Nuʻuanu Avenue side is a winding stairwell. Today the station looks as it did in 1931, although its tenants have changed from law enforcement and court officials to property tax and housing employees. In 1991 the building was named the Walter Murray Gibson Building as a tribute to one of the most colorful and perhaps misunderstood figures in nineteenth-century Hawaiian politics. (See Historical Notes)

• *See Historical Notes (page 124) for more information.*

The old Honolulu Police Station, now called the Walter Murray Gibson building, is located between Nu'uanu Avenue and Bethel Street, at 842 Merchant Street. Erected in 1931, it is the second police station to stand on this site. The building is considered one of the city's architectural treasures and is among the last of the Mediterranean-style structures built in the downtown area.

Punahou School

The first missionaries to Hawai'i endured unending sacrifices to establish a mission and educate the people of Hawai'i. While they labored to teach the local natives, their own children were sent back to New England for a more formal education. But eventually it became imperative to provide for their children's education in Hawai'i. As early as 1832, the misssionaries met to discuss the founding of a boarding school on O'ahu for missionary children from all the islands.

The Punahou land, known as Ka Punahou, "The New Spring," had been the property of Hoapili, who gave it to his daughter, Liliha, wife of Boki, the governor of O'ahu. Ka'ahumanu, the queen regent, became an ardent supporter of the missionaries and wished to give them land. She consulted Hoapili, who suggested Ka Punahou, even though he had already given it to Liliha. Despite Liliha's protestations, the land was given to Hiram Bingham and, according to missionary rules, became the common property of the whole mission. Punahou School's first building, an adobe structure with thatched roof, opened in 1842 for fifteen pupils, ranging in age from five to twelve. Rev. Daniel Dole, Punahou's first principal, promoted classical learning with vocational training, so that pupils would be prepared to assume useful occupations in the community if they chose to remain in the islands. The school grew its

Punahou School, located at 1601 Punahou Street, was founded in 1841 by Congregational missionaries for the education of their children. It continues to be one of Hawai'i's foremost private schools, with 3,700 students, kindergarten to grade 12. The building depicted here, Old School Hall, built in 1851, is the oldest structure on the campus and is still used today for classes.

75

PUNAHOU SCHOOL

own vegetables, and boys were required to devote time to cultivating the land.

By 1844 it had become apparent that Punahou School would have to expand to accommodate the increasing number of students. After supplications to the American Board of Commissioners for Foreign Missions to fund a new building, Dole Hall was completed in 1849. In the same year, the Hawaiian government granted the institution a charter in which it was officially designated Ka Punahou School.

Dole Hall, a two-story coral stone structure, provided lodging downstairs for the Dole family and a girls dormitory upstairs. By this time, William Harrison Rice had been summoned from Maui to superintend the work on the school's farm. After ten years of dedicated labor, Rice left Punahou to manage the Lihue Plantation on Kaua'i. Before he left in 1851, he saw that the Old School Hall was built, today the oldest building on the Punahou campus. This simple, rectangular, two-story, stone building has a coral foundation laid directly on the ground and stone walls covered with plaster and white paint. Originally the building had verandas built on both levels.

By the late 1840s, most of the older students were being sent to college in the United States. Many missionaries chose this time to leave the mission and accompany their children to settle once again in their homeland. The Foreign Missions commissioners grew concerned that the Sandwich Islands mission was in jeopardy, as more and more missionaries abandoned their posts. Late in 1849 they discussed the possibility of founding a college in Hawai'i. A charter was drawn up in 1853 and the corporate body known as "The Trustees of the Punahou School and O'ahu College" was created. In 1854 the name was shortened to "The Trustees of O'ahu College," and Edward G. Beckwith, a twenty-seven-year-old graduate of Williams College, Massachusetts, was appointed president, having served as principal of the Royal School in Honolulu. In 1934 the name of the school was officially changed back to Punahou; today, as a reminder of the college, the letter "O" is awarded to athletes.

Though the face of the Punahou campus has changed over the years and modern buildings have replaced much of the adobe, coral, and thatched roofs of earlier days, Old School Hall remains a landmark that is still used as a classroom building. The spring, Ka Punahou, provides water for the lily pond and the entire campus. Pauahi Hall, built in 1898, retains its original stone finish and unique architectural design. Cooke Hall is now used for faculty and administration. All the buildings are living memorials to the early work of the missionaries and their influence on education in the islands and to the contributions of more recent faculty and supporters of Punahou.

• *See Historical Notes (page 124) for more information.*

Queen Emma Summer Palace

Hanaiakamalama, meaning "Foster Child of the God Kalama," was one of the two summer palaces used by Hawai'i's royal families. The other is Hulihe'e Palace, in Kailua-Kona on the Big Island. High Chief Keoni Ana, otherwise known as John Young II, named the house after one of the ancestral gods of his mother, a high chiefess and niece of Kamehameha I. It was the name of their home in Kawaihae. Young was the premier of Hawai'i during the reigns of Kamehameha III and Kamehameha IV. His father, the famous Englishman John Young I, was a trusted adviser and friend to King Kamehameha I. Near the site of the summer palace, he and Isaac Davis held council with Kamehameha I and his generals on the day of battle in 1795, just before the victorious drive that pushed Oahuans to the edge and sent them hurtling over Nu'uanu Pali.

When Emma Rooke married King Kamehameha IV in 1856, her uncle, John Young II, made them a gift of the use of Hanaiakamalama; when he died in 1857, the property was left to Emma, who was then queen. She and her husband held court informally here and enjoyed relaxation from the rigid formalities of life at the first 'Iolani Palace. It was a favorite place of retreat, a place where the young Prince Albert Edward could frolic in freedom in its beautiful gardens and alongside a pond speckled with pink water lilies and traversed by a rustic arched bridge.

The summer palace was a white frame house, modest in design and comfortably adapted to the Nu'uanu Valley's frequent showers and cool breezes. Its wide central hall, high ceilings, and floor-length shuttered French doors on all sides made it airy and seemingly spacious. Each room had access to the lanais and gardens outside, where beautiful flowers and shrubs formed a charming exterior to complement the Victorian decor inside. The front lanai was supported by six Doric columns, and a ballroom ran the width of the building.

No records concerning the construction of the house exist. It is believed to have been constructed in Boston and then shipped to Hawai'i via Cape Horn. The architect may have been Charles W. Vincent, one of the few skilled builders in Hawai'i at the time.

Queen Emma was a devoted member of the Episcopal church, and it was through her efforts that the first Anglican church (later called St. Andrew's Cathedral) and its mission school, the Priory, were established. When Emma died in 1885, her will stipulated that any income from the property would go toward scholarships for girls at St. Andrew's Priory. Unfortunately, her estate was so deeply in debt that Hanaiakamalama was put up for auction in 1886. The Spencer family occupied it from 1911 to 1913, and then the territorial government decided to use the land to build a filtration plant for Nu'uanu Reservoir and a public park. The already deteriorating house was to be demolished. At this time, the Daughters of Hawai'i, a society dedicated to the preservation of Hawaiian culture, intervened to rescue and restore the building and convert it to a museum. Extensive repairs, largely of termite damage, were required. The Daughters of Hawai'i also sought to recover the personal possessions of Queen Emma and original furnishings from the period when Hanaiakamalama served as a royal residence. Many of these are on display in the museum. Hanaiakamalama preserves the many mementos of a queen whose love and generosity deeply touched and uplifted her people.

• *See Historical Notes (page 125) for more information.*

Following page: **The summer palace of Queen Emma and King Kamehameha IV is located at 2913 Pali Highway. This small, cottage-like residence is now a museum containing rare artifacts and personal belongings of the royal family. Nu'uanu Valley affords a lovely setting for what was once a favorite retreat of the king and queen. The site is open every day of the week, 9:00 AM–4:00 PM.**

81

RICHARDS STREET YWCA

RICHARDS STREET YWCA

In downtown Honolulu, directly across the street from ʻIolani Palace and the coronation pavilion, stands the attractive Richards Street YWCA. Its simple, harmonious elements are characteristic of the designs of Julia Morgan, one of America's most talented and well-known architects. Morgan is credited with over one thousand buildings during her fifty-three year career, the most renowned being Hearst Castle in San Simeon, California. Her portfolio includes an unusually diverse assortment of subjects. Despite her world renown, she was a private woman who did not seek public acclaim, shunned the press, and secretly disposed of all her office records before she died in 1957.

Morgan first visited Hawaiʻi in 1917 and remained long enough to remodel a Waikiki Beach house for the Atherton family. Later, when local YWCA officers began planning to build a new administration building, they chose Morgan as the architect and Catherine Jones Richards as landscape designer. After receiving this commission, Morgan made one trip to Hawaiʻi to study the site and develop the plans. She did not personally oversee the project but sent her assistant, Edward Hussey, who made frequent trips to Honolulu.

The building was completed in 1927 at something under $400,000. In describing the Richards Street YWCA to a former director of publicity, Morgan referred to it as "unusually frank and sincere architecturally." She noted that there was no "false work" or "furring" in the building and that all parts of the structure, whether girders, beams, or arches, were given a decorative quality. Although it was not easy making a reinforced concrete building without the usual "softenings" such as wainscots, cornices, trims, and paneling, still the final result was a building that she called "friendly, cheerful, and interesting." Even though its dimensions were large, it escaped appearing "clumsy, monotonous, or gloomy." Morgan counted the YWCA among her eight most favorite projects.

The building consists of three main units tied together by double-decked loggias. The main wing houses the general reception rooms, offices, clubs, and classrooms. The second wing is separated from the main wing by a double court. The third wing, a two-story building fitted between the two main wings, houses the locker rooms and dressing rooms. There are plenty of open spaces and windows, allowing trade winds to blow through, cooling and freshening the halls, corridors, and open courtyard. The swimming pool in the minor courtyard has been the scene of many a swim class or fun-filled session of free swim for adults and children.

There is an atmosphere of charm and a hum of activity about the building which seems to be intrinsic to its design and the purpose for which it was built. Perhaps of all the historic buildings in Hawaiʻi, it is the most utilized, most enjoyed, and most intimately involved in the lives of the people.

Previous page:
Across from ʻIolani Palace, at 1040 Richards Street, stands a remarkable building often taken for granted in its utilitarian YWCA role. Aside from its attractive appearance, the structure is all the more distinguished as a product of one of America's foremost female architects, Julia Morgan.

Royal Brewery

Just *makai* of the Kawaiaha'o Church property, in an area known as Kaka'hawaiiako, stands the original home of Primo Beer and Royal Beer. This red-brick building is the only remaining structure of what was once a complex brewery plant. It was built as an industrial building, with a great deal of attention paid to the external appearance. The Queen Street facade is of grand proportion, and its three-story brick arches, corbels, and intricate patterns make a rare architectural statement.

The name on the front facade reads "The Honolulu Brew'g & Malt'g Co." As the original home of Primo Beer, the Royal Brewery is believed to be the oldest brewery in Hawai'i, as well as the only building of its kind in the islands. Built in 1900, only two years after the annexation of Hawai'i by the United States, the brewery operated into the late 1960s. It was owned by American Brewing Co., Ltd., a subsidiary of C. Q. Yee Hop & Co., Ltd.

The imposing brick and steel edifice was built of materials that came by sailing ships from San Francisco and New York via Cape Horn at the turn of the century. The four-story, 80-foot-high structure, with its big brick arches and intricate wrought-iron patterns, reflects the finest masonry craftsmanship of the gaslight era. The brewery includes a rathskeller, a German-style beer hall, and a mural depicting Rip Van Winkle waking from his long sleep. The old plant was designed to specifications of the Honolulu Brewing and Malting Company by a New York architect who specialized in brewery construction. One of its brewmasters, William Glockner of a well-known brewing family of Munich, Germany, emigrated here to run the plant.

The structure stands on a 77,000-square-foot lot and is listed on both the State and National Registers of Historic Buildings. Originally there were three buildings, including a two-story bottling house behind the main building and an ice-making plant. From 1961 until 1993, the one remaining building stood abandoned and demolition seemed imminent. What saved the brewery was the possibility that it could be restored as a heritage house—a Hawaiian history center and orientation point for both visitors and local residents. Other ideas for the structure have been an opera house, a children's theater, and a theater for other performing arts groups.

At this time the Hawai'i Community Development Authority is planning to use the red-brick building to develop office space and a senior citizens' community center. Adjoining complexes will be built for seniors' furnished rental apartments and affordable condominiums.

Following page:
An interesting and unusual sight at 553 South Queen Street, the old Honolulu Brewing and Malting Company, also known as the Royal Brewery, was the original home of Primo Beer and Royal Beer. It is presently abandoned and awaiting plans to develop the site for offices and housing.

Royal Hawaiian Hotel

Set in a garden of tropical shrubbery with coconut palms swaying in the sea-cooled breezes, the Royal Hawaiian Hotel is undoubtedly one of Hawai'i's most renowned landmarks. Built in 1927, it occupies a site once known as the Helumoa coconut grove, where King Kamehameha V enjoyed leisure times at his beach cottage. Early records tell us that this area was once a *heiau pookanaka,* or sacrificial temple, strictly *kapu* to commoners. From the many traditional game stones discovered here during excavation, it appears that the land was used at one time as a sports field. Long before Kamehameha and his armies invaded the island of O'ahu, the site seems to have been a recreation site favored by the kings of O'ahu and Maui.

The Royal Hawaiian Hotel was developed and financed by the Matson Navigation and Territorial Company. It was designed by the New York architectural firm of Warren and Wetmore, a nationally reputed company credited with such famous hotels as the New York Ritz-Carlton, Belmont, Vanderbilt, Biltmore, Ambassador, and Commodore. In a style described by architects as Spanish Mission Revival, Spanish-Baroque, or Spanish-Baroque-Moorish, it was a striking departure from contemporary Hawaiian architecture. Amid lush foliage and coconut trees, the pink plaster walls, scalloped parapets, blue tile roofs, ornate bell tower, terra cotta planters, and domed cupolas brought to mind a fairy tale castle materialized on the white sands of Waikiki Beach. Originally a site encompassing twenty-one acres of landscaped grounds, the hotel and grounds now occupy less than ten acres.

The Royal Hawaiian opened on February 1, 1927, and was an immediate favorite among visitors. The *Advertiser* described it as "a coral pink castle set amidst a royal grove of old Hawai'i." It featured a theater and auditorium with two stages, a ballroom, and a banquet hall. At a cost of $14 per day, the room rate was higher than any other hotel in Hawai'i. The Moana was charging $8 per day and most other hotels charged $5 or less.

The Royal Hawaiian Hotel is unique in many ways. The original building had sections of four and six stories, with towers at various locations. Its main entrance, on the sea side of the building, features a beautifully decorated porte cochere, with Tuscan columns and buttressed piers leading into a richly decorated lobby. Four square towers with round, arched openings are topped with blue tile hipped roofs. Expansive gardens, with monkey pod, banyan, and coconut trees, once surrounded the hotel. Certain of the trees are believed to predate King Kamehameha V's time here. The hotel's open balconies, with bamboo awnings, wrought iron handrails, and terra cotta fronts with double doorways, grace the exterior, while open archways and arcades allow gentle trade winds to cool the interior.

Today the Royal Hawaiian, though overshadowed by its high-rise neighbors, is still admired by all who come to Honolulu. Despite occasional rumors of replacing the hotel with a modern, larger building, it remains a charming oasis on the shores of Waikiki.

• *See Historical Notes (page 125) for more information.*

Waikiki's "Pink Palace" is one of Hawai'i's best-loved luxury hotels. Although practically hidden by neighboring high-rise buildings at its 2295 Kalakaua Avenue address, the Royal Hawaiian continues to draw visitors who marvel as its lush tropical garden setting and charming atmosphere. The Royal Hawaiian, built in 1927, is one of the oldest remaining hotels in Waikiki.

ROYAL HAWAIIAN HOTEL

Royal Mausoleum

In Nuʻuanu Valley, once a favorite place of retreat for Hawaiʻi's royal families, stands a picturesque chapel with neighboring crypts in which most of Hawaiʻi's royalty and other significant personages of the kingdom were laid to rest. The Kamehamehas, Kalakauas, and friends of the royal families such as T. C. B. Rooke, John Young, his daughter Jane, and R. C. Wyllie are entombed here. Before they were placed in separate crypts, the caskets of the royalty were stored in the chapel.

The area in which the Royal Mausoleum, also known as Mauna ʻAla, is located is as historic as the monument itself. It is said that the Battle of Nuʻuanu, in which King Kamehameha I vanquished the armies of King Kalanikupule in 1795 and established himself as ruler over the Hawaiian Islands, was fought here.

The original royal mausoleum was located on the grounds of ʻIolani Palace, in the area once known as Pohukaina. When the British frigate *Blonde* arrived in Honolulu in 1825 bearing the bodies of King Kamehameha II and Queen Kamamalu, who had died in England in 1824, the first mausoleum was constructed to receive their remains. It served as the royal burial place until 1865. After the death of his infant son, Prince Albert Edward, the grieving King Kamehameha IV ordered a new and larger mausoleum to replace the already overcrowded one. Construction began in 1863 after the king's death and, upon its completion, he and his son were the first to be buried there. With the exception of King Kamehameha the Great—who, according to ancient tradition, was buried secretly—and King Lunalilo—who chose to be buried apart from the Kamehamehas—all of Hawaiʻi's royalty are entombed in Nuʻuanu Valley.

Theodore Heuck designed the Royal Mausoleum to resemble Gothic tombs in Europe. The coral block building is in the shape of a Greek cross with stained glass windows. When the chapel was completed in the fall of 1865, the contents of the old mausoleum were transferred to their new resting place in what was described by news reporters of the time as "an eerie, utterly pagan ceremony." A solemn procession started at nine o'clock in the evening, with women wailing and *kahunas* chanting, by the flickering light of kukui torches, as eighteen caskets were carried over the distance of two miles from the palace grounds to Nuʻuanu Valley. The procession of chiefs and members of the royal family was headed by King Kamehameha V and his father, the aged Governor Kekuanaoʻa. Upon arriving at the site, the caskets were placed inside the mausoleum, in order of lineage, on koa wood stands, and Christian burial rites were conducted.

After David Kalakaua became king, he had the remains of his father, mother, and an infant son removed from Kawaiahaʻo cemetery to the Royal Mausoleum. In time, his remains and the remaining members of his family were interred there, too.

In 1887 the Kamehameha dynasty caskets were removed from the mausoleum to a private vault, followed in 1904 by the Kalakaua dynasty. The building that once served as mausoleum for the great families of Hawaiʻi was converted to a chapel in 1922 and continues to be called the Royal Mausoleum. Its koa wood interior has been restored, and on its altar are inscribed the words "Hemolele, Hemolele, Hemolele" ("Holy, Holy, Holy").

An ornate, gold-tipped wrought-iron fence surrounds the Royal Mausoleum, obscuring it from traffic passing by its 2261 Nu'uanu Avenue location. The chapel depicted here was originally built as a mausoleum in which the remains of Hawaiian royalty and their families were kept. The grounds are open Monday through Friday, 8:00 AM–4:30 PM.

St. Andrew's Cathedral and Priory

Queen Emma (1836–1885) and her husband, Alexander Liholiho, King Kamehameha IV, brought the Episcopal church to Hawai'i and shared in the founding of St. Andrew's Cathedral and Priory School for Girls. Kamehameha IV came to the throne in 1854 and was familiar with Anglican services, having attended them in both England and the United States while touring those countries some years earlier. He had grown tired of the puritanical American missionaries and felt Hawai'i needed a change from the religion that had infiltrated the internal affairs of his country.

King Kamehameha and Queen Emma solicited funds for the project from friends at home and abroad. The king offered land for a church and parsonage and a gift of 200 pounds annually to the missionary group that would be willing to come to Hawai'i. Queen Emma went about personally collecting donations. In 1861 the Reverend Dr. Thomas Nettleship Staley was consecrated the first bishop of the Honolulu diocese, having come to the islands at the king's request in 1862. Queen Emma was baptized soon after his arrival, and her name was the first one entered in the cathedral register, even before a church was started. That same year her only son, Prince Albert Edward, died, and one year later her husband died. The bereaved Queen Emma mourned for almost two years before she resumed her efforts at raising funds for the church. She sailed to England in 1865 to seek financial assistance from Episcopal parishioners there. Her friendship with Queen Victoria had been long established, and Queen Emma was warmly received at Windsor Castle. She remained in England for six months and raised $30,000. There she commissioned London architects to design her cathedral, and she paid for blocks of sandstone cut from quarries in England to be shipped as ballast aboard vessels bound for Honolulu via Cape Horn. Her dream was realized on March 5, 1867, when King Kamehameha V laid the cornerstone of St. Andrew's Cathedral. He dedicated it to the memory of his younger brother, Kamehameha IV, who had died on the feast day of Saint Andrew. Soon after, Mother Superior Priscilla Lydia Sellon of the Sisters of the Holy Trinity, a teaching order of the Anglican church, came to Hawai'i with a group of sisters. In three weeks the buildings of St. Andrew's Priory School for Girls were begun.

Construction of the cathedral proceeded slowly, and, when Bishop Staley returned to England in 1870, it stopped for twelve years. Ninety-one years passed before the cathedral was completely finished. The consecration took place in 1902, when the first two bays were finished. During that year it was officially received into the Protestant Episcopal Church of the United States, since Hawai'i had been annexed to the United States in 1898.

By 1908 two more bays were added, and by 1912 the Alice Mackintosh Memorial Tower was completed. In this great tower a twenty-five-bell carillon was placed. Finally, in 1958 the cathedral was finished with the addition of the nave's last two bays, the narthex, vestibules, and the Great West Window. This brilliant stained-glass mural depicting many events in the history of the Christian church from Judea to Hawai'i, was designed in 1956 by John Wallis of Pasadena, California. In front of the cathedral stands a fountain of Saint Andrew, surrounded by ten bronze sculptured fish spouting streams of water from their mouths. The raised curbing around the fountain is inscribed with the words "Go ye into all the world and preach the Gospel to every creature." The fountain was designed by Carleton Winslow of Beverly Hills, California, and executed by Mario Valdastri & Son of Honolulu. The figure of Saint Andrew was created by the renowned Ivan Mestrovic.

St. Andrew's Cathedral is a remarkable accomplishment, with its massive columns and Gothic vaulting leading to a beautiful open chancel with pierced stone apsidal screens. The baptismal font, one of its oldest and most treasured gifts, was sent from England by Lady Jane Franklin in 1862 for the baptism of Prince Albert Edward. It is carved from marble and Caen stone, a naturally tinted pinkish and yellowish stone from France. On its front in bas relief is a scene of the baptism of Jesus by Saint John. Unfortunately, the baptismal font arrived too late for the young prince, who had passed away suddenly in 1862. The pulpit, communion rail, lectern, and the exquisitely carved high altar are all made of carved Caen stone.

• *See Historical Notes (page 126) for more information.*

The Episcopal church was established in Hawai'i through the efforts of King Kamehameha IV and Queen Emma. The name St. Andrew's was given to its first church because it was on this saint's feast day that the king died. St. Andrew's and the Priory School for Girls are located at the corner of Beretania and Queen Emma streets. The cathedral is noted for its beautiful gothic style, bell tower, and stained-glass windows.

Ali'iolani Hale
The Judiciary Building
Honolulu, Hawaii

The statue of King Kamehameha I stands in the center of a circular lawn in front of Ali'iolani Hale, Honolulu's State judiciary building at 417 South King Street. It is one of the most popular tourist attractions on O'ahu. Ali'iolani Hale's Judiciary History Center is open to the public. For information on tours and hours of operation, call (808) 548-3163.

Statue of King Kamehameha I and Ali'iolani Hale

In the legislature of 1878, Walter Murray Gibson proposed the idea of a monument to the centennial of Hawai'i's "discovery" by Captain James Cook. Since Kamehameha the Great was among the first to greet the discoverer in 1778, and was renowned for having unified the islands under one rule, he was chosen to be immortalized in bronze. The legislature appropriated $10,000, and Boston sculptor Thomas R. Gould received the commission. Today, in the heart of Honolulu's historic district, stands this famous Hawaiian landmark.

Gould's first problem was how to make a credible statue of the king, when his only reference was an engraved portrait of Kamehameha I by artist Louis Choris, which had to be modified to portray a younger man of about age forty-five. Gould studied photographs of natives in traditional royal attire, with feather cloak, baldric, helmet, and spear, and patterned the king's body after that of Robert Hoapili Baker, who was considered to have the correct type of physique. Gould worked at his studio in Florence, Italy, and, when completed, the mold was shipped to Paris and cast in bronze. The statue was exhibited briefly in Paris, where artists, diplomats, and French citizens thronged to gaze upon the hero of a strange and remote land.

Meanwhile, in Hawai'i, discussion centered around choosing an appropriate resting place for the monument. The legislature favored the grounds in front of Ali'iolani Hale and agreed to appropriate $2,000 for a pedestal and protective railing. Gould was also commissioned to produce four bronze tablets depicting scenes from the king's life: his visit aboard Captain Cook's ship, a battle scene in which he warded off five spears hurled at him, his fleet of war canoes off the coast of Kohala, and the era of peace promulgated under the edict referred to as "The Law of the Splintered Paddle."

The statue was shipped from France in August 1880 and was expected in Hawai'i in mid-December. In February 1881, word came that the bark *G.F. Haendel* had gone down off the coast of the Falkland Islands with all cargo lost. A replica of the statue was ordered at a cost of $7,000 and was expected to arrive sometime in July or August 1882. But in March the British ship *Earl of Dalhousie* arrived in Honolulu Harbor, and rumor that the Kamehameha statue was aboard swept the town. It was not the replica but the original statue, which Captain Jervis had found in front of a store in Port Stanley in the Falkland Islands, bought, and put aboard his ship en route to the Hawaiian Islands. Gibson, chairman of the monument committee, hurried to Jervis's ship and paid him $875 for the recovered statue. More than thirty years later, the story was told that, after the *Haendel* had sunk, a fisherman sighted the remains of the lost ship and the box containing the statue. The Kamehameha statue was recognized and removed to Port Stanley—with its right hand broken off, spear damaged, and a hole in the feather cape. The finders of the treasure placed it as a guardian over the little port.

While the original statue underwent repairs in a small shed near Ali'iolani Hale, questions arose about what to do with its replacement, which was expected to arrive in time for unveiling at the belated coronation of King Kalakaua. The British ship *Aberaman* arrived in Honolulu in January 1883, two weeks before the coronation, carrying the replica, the bronze tablets, and a forearm for the damaged original statue. The replica was set in place and covered with a royal standard and a Hawaiian flag. On the day of Kalakaua's coronation, an enthusiastic crowd cheered as their king parted the two flags to reveal the striking figure of their hero. With spear in left hand and right hand outstretched with open palm, the great warrior seemed to beckon to the people for whom he had fought to establish peace and unity. After the unveiling, Hawaiians lingered for hours, softly chanting and offering homage to their great king.

When statehood was granted in 1959, Hawai'i was given the customary privilege of selecting two statues to be reproduced for Statuary Hall in the U.S. Capitol. The chosen statues were of Kamehameha the Great and Father Damien, the leper priest of Molokai. Both statues received much acclaim, but the statue of Kamehameha merited distinction for being the largest in the entire collection of ninety statues, the only statue whose unveiling was announced with the blowing of a conch shell, and the only one to be adorned with leis. This was the first statue of a king to be included in Statuary Hall.

The statue of Kamehameha the Great continues to be admired as the finest monument to a Hawaiian ruler and a lasting symbol of the strength and dignity of the Hawaiian monarchy. It is one of Hawai'i's most visited landmarks. Each year on June 11, a state holiday, it is adorned with dozens of 30-foot leis to honor the great king who unified the Hawaiian Islands.

Behind the Kamehameha statue stands Ali'iolani Hale—the Judiciary Building. Originally intended to be a royal palace, Ali'iolani Hale served as the seat of the legislature and cabinet offices of the monarchy.

King Kamehameha V commissioned Thomas Rowe of Sydney, Australia, to design his palace. By the time the king received the plans, he had decided that a government office building would be more practical. Thus the original plans were modified by the superintendent of public works, Robert Stirling. In an impressive Masonic ceremony on February 19, 1872, Kamehameha laid the cornerstone for the Judiciary Building.

Kamehameha V died shortly afterwards, and in 1874 King Kalakaua formally named the building Ali'iolani Hale, "House of the Heavenly King," in his honor. "Ali'iolani" was one of the sacred names given to Kamehameha V at birth. Unlike his predecessors, King Kalakaua did not wish to devote such a grand structure strictly to the mundane needs of government business. By day it served government purposes, but by night it became the scene of festive balls and royal entertainment. It was the only building in Hawai'i suitable for such entertainment: the future home of King Kalakaua—'Iolani Palace—was yet to be built.

Ali'iolani Hale was constructed of cast concrete block, a material first used for the Kamehameha V Post Office. Hawaiians were adept at laying up rough or broken stones after centuries of building *heiau* and other structures from lava rock. To assist them with the concrete block construction, two stone masons, Lishman and Quinton, were brought in from Sydney. For months molds were made, concrete mixed, and blocks, columns, and balustrades cast and set in place. One of the finishing touches was a huge clock with four dials, manufactured in Boston for an exorbitant $400. The building's neoclassic exterior with Italian-piazzo style clock tower has changed little since the building was completed in 1874, at a cost of $130,000.

Ali'iolani Hale contained Hawai'i's first public library, the National Museum, and the Natural History and Microscopical Society, an organization sponsored by King Kalakaua. Most of its treasures, relics of the Kamehameha dynasty, were removed in 1890 to the Bernice Pauahi Bishop Museum in Kalihi.

Today, Ali'iolani Hale is home to the Hawai'i Supreme Court and the Hawaiian Judiciary History Center museum, located on its ground floor. This educational, nonprofit facility, opened in 1989, displays and interprets Hawai'i's judicial history. It serves as a center of research and preservation of information and artifacts belonging to the Hawaiian judicial system.

• *See Historical Notes (page 126) for more information.*

Territorial Building

The Territorial Building, built in 1926, is one of the few historic buildings in Honolulu designed and constructed by a local architect, Arthur Reynolds, and a local contractor, Alfred H. Olund. Reynolds was assisted on the project by C. B. Ripley. All the labor and materials for the building were from local sources, as well.

Reynolds designed the building to meet a modern challenge: shortage of working capital. The territorial legislature had appropriated only $250,000 for the building's first stage, and there was no way of knowing what it might appropriate in its next meeting. Reynolds began with a one-story building, with a structure sufficient to support a central five-story tower for additional offices. As it turned out, a second appropriation of $250,000 came through and a central tower portion was added with fluted Corinthian columns in the arch above the entry. This composite approach met with skepticism from other architects, who thought the tower appeared awkward and did not contribute to a unified structural design.

The Territorial Building's exterior is of stuccoed reinforced concrete. Unlike its neighbors, which introduced Mediterranean and Italian Renaissance styles into the Capitol District, this structure brought a new Classic Revival style, simple in design yet stately and functional. It had plenty of windows to allow for natural ventilation and lighting. Even the cut-glass dome above the lobby was once lighted by a shaft to the roof. This dome was not in the original plans but was inspired by Lyman H. Bigelow, the superintendent of public works. It featured the Hawai'i coat of arms surrounded by American flags. The area was later filled in with offices, and today the dome is artificially illuminated. During a 1975 restoration project, architect Herbert Matsumura discovered that the dome's glass panes had been wrongly placed after a cleaning and now spelled out "awaii ry of H Territori." This had been a source of great amusement to visitors for many years. Matsumura had the panes put back correctly to read "Territory of Hawai'i."

The original building had a fire escape that extended down only as far as the roof of a covered lanai; from that level, there was no way to the ground except by jumping.

The Territorial Building was used to house territorial and later state offices until the new state office building was erected. When it was rededicated in 1978, it was renamed after Mataio K. Kekuanao'a, in honor of the father of Kamehameha IV and Kamehameha V. It was placed on the State Register of Historic Places in 1975 and on the National Register in 1978.

TERRITORIAL OFFICE BUILDING
Honolulu, Hawaii

Honolulu's Territorial Building, located at 465 South King Street, stands out as an unusually designed structure. Unlike most of its Italian Renaissance neighbors, it introduced a Classic Revival style to the historic district. The building was originally used for territorial, then state, offices; today it houses the legislative auditor, ombudsman, Public Utilities Commission, and State I.D. office.

Thomas Square

In February 1843, the British warship *Carysfort* arrived in Honolulu Harbor to address the complaints of Englishman Richard Charlton, who felt that British subjects were being mistreated by the Hawaiian government. The commander, Lord George Paulet, presented a series of demands to King Kamehameha III and promised to attack if the king did not comply. A desperate and defenseless Kamehameha ceded the islands provisionally to Great Britain, and the British flag was raised over the fort in Honolulu. For five months, foreign rule under Paulet was enforced over the Hawaiian Islands.

Relief came in July 1843, when the ship carrying Rear Adm. Richard Thomas landed at Honolulu Harbor. Outraged at what had transpired during the brief time Paulet had been in the islands, Thomas declared that the cession was not approved by the British crown and repudiated Paulet's actions. On July 31, in an elaborate ceremony held in a plains area east of town, foreigners and natives gathered by the thousands to hear the address by Admiral Thomas, who arrived in the king's state carriage. The king and his entourage arrived on horseback and were greeted respectfully with a twenty-one-gun salute. At Thomas' signal, the British flag was lowered and the Hawaiian flag hoisted in its place amid salvos of guns from all the ships in Honolulu Harbor.

A prolonged and mighty cheer arose from the crowd. Admiral Thomas read a lengthy declaration that restored sovereignty to the islands. At the close of the ceremonies, the crowd dispersed and reassembled at Kawaiaha'o Church to hear Admiral Thomas' speech translated into Hawaiian. This time the king addressed his people joyously, speaking the words that would become Hawai'i's motto: *Ua mau ke ea o ka aina i ka pono*—"The life of the land is preserved in righteousness."

Next followed a general holiday of ten days, during which festivities never lagged. An unsurpassed luau was given by the king in Nu'uanu and exhibitions of ancient Hawaiian games were held. It was a time of jubilation for all the people of Hawai'i, except some disgruntled Britons.

The area of land on which the restoration ceremonies took place was named Thomas Square in honor of the admiral. Today it is a park with banyan trees skirting a central fountain and shade trees around its perimeter. Many of O'ahu's craft fairs and other cultural events are held in this lovely setting. Thomas Square is distinguished as a historic landmark commemorating Hawai'i's struggle to remain an independent and sovereign nation in the face of encroaching foreign interest.

• *See Historical Notes (page 127) for more information.*

Not far from downtown Honolulu is a historic park bounded by South King, Beretania, and Victoria streets and Ward Avenue. Named after Rear Adm. Richard Thomas in 1843, Thomas Square was the site of the restoration of Hawaiian sovereignty after a brief takeover by Great Britain. Its beautiful banyan trees and central fountain make it a delightful gathering place and scene of cultural events and craft fairs.

U.S. Post Office, Custom House and Court House

$\mathcal{A}$T THE TURN OF THE CENTURY Honolulu was the principal port of entry into Hawai'i. Customs collections had begun during the reign of King Kamehameha I, when chiefs paddled out to ships anchored offshore to collect fees on the goods they carried. The first customhouse was built in 1848, near the waterfront on Queen Street between Nu'uanu Avenue and Smith Street. It was called the "Custom House in the Pacific." The second customhouse, completed in 1860, was on the corner of Fort and Allen streets.

The early 1900s marked a period of rapid development in Honolulu. As trade and commerce flourished, more space was needed to house the expanding customs offices and other government agencies. Since the overthrow of the monarchy in 1893, 'Iolani Palace had been occupied by government offices and was overcrowded. Plans for a new federal building were drawn up by architects York and Sawyer of New York by 1910. It took until 1916 for all the pieces of property in a four-acre lot near 'Iolani Palace and Ali'iolani Hale to be purchased. The old opera house on the corner nearest Ali'iolani Hale had been demolished, leaving that space available as well. Construction was delayed for many years by budgetary problems and the involvement of the United States in World War I. By 1918 the original plans had to be modified to conform to a new budget, and construction finally began in 1921. The $1.5 million Federal

The U.S. Post Office, Custom House and Court House, often referred to as the Old Federal Building, occupies the block between Richards and Mililani streets at 335 South King Street. It contains a post office, customs offices, and state agencies.

• *See Historical Notes (page 127) for more information.*

Building, as it was commonly called, was completed on April 1, 1922. It was officially named the "United States Post Office, Custom House and Court House," but the name was not affixed to the structure until 1955. York and Sawyer were assisted by local architects Walter Leavitte Emory and Marshall Hickman Webb. The style they used for this government building has been referred to as both Spanish Colonial Revival and Mediterranean, bearing elements in common with other downtown buildings such as Honolulu Hale, the YWCA, the Hawaiian Electric Company, and the Armed Services YMCA. The architects' endeavor to create a structure compatible with the island ambiance without the typical appearance of a ponderous government office building proved successful.

The Federal Building has a U-shape, with three floors and an open courtyard and spacious porticoes in front. Customs offices on the second floor were designed with counters and had wainscoting of Italian marble. The central portion contained the courtrooms on the upper floors. The post office quarters on the west side included intricate passageways and secret louvers for the supervision of mail handling. Originally there were arched wrought-iron gates on the Richards Street side, but these were removed in the 1960s. The windows facing the courtyard were arched and covered with decorative window grills. Two marble staircases led to the second floor. The building was constructed with a full basement and two square towers rising to six floors. A three-story addition on the Mililani Street side was designed in 1929 by the treasury department. It assimilated the same style as the original building and was completed in 1930. The addition contained two courtyards, one with access to Mililani Street, and the other opening toward Richards Street. The architectural character of the exterior, with its thick plaster walls, arched openings, and deep overhanging tile roofs, has been retained in spite of minor changes over the years.

In 1979 the U.S. Postal Service purchased the building from the Hawai'i General Services Administration for approximately $8 million, and a major renovation project followed. The building is now occupied by the post office, offices of the governor and lieutenant governor, the Judiciary Intermediate Court of Appeals, the Department of Hawaiian Home Lands, the U.S. Customs Service, the State Foundation on Culture and the Arts, and other state agencies.

USS ARIZONA MEMORIAL

WHEN JAPANESE TORPEDO BOMBERS moved into attack formation as they neared Pearl Harbor, the USS *Arizona* was tied up at Quay F-7, alongside the repair ship *Vestal*. The pilots could easily see that the Pacific Fleet had been caught by surprise, which they heralded with the code words "Tora! Tora! Tora!" As the bombs began to drop, the *Arizona*'s air raid alarm sounded, and moments later the first bomb struck astern. Two more bombs exploded on the aft quarterdeck. One armor-piercing bomb penetrated the forecastle and exploded in one of the ship's forward magazines. The resulting fires and explosions wrecked the forward part of the ship, killing hundreds of men, including the *Arizona*'s commanding officer, Capt. Franklin Van Valkenburgh, and the battleship division commander, Rear Adm. Isaac C. Kidd.

So effective were some of the direct hits that the fire-engulfed *Arizona* sank in about nine minutes, trapping hundreds of men below. The *Arizona* lost over 70 percent of her complement: 1,177 officers and men killed or missing. Only 75 bodies were recovered; the remaining men were entombed in the sunken battleship. This was December 7, 1941—the day the United States and the Territory of Hawai'i went to war.

The *Arizona* was named for the forty-eighth state to join the Union. The Pennsylvania-class battleship was built in the Brooklyn Navy Yard, and the keel was laid during ceremonies on March 16, 1914. The *Arizona*'s design included twelve 14-inch guns in four triple turrets, two forward and two aft, capable of projecting 1,400-pound armor-piercing shells a maximum of 18,000 yards. The secondary battery of twenty-one high-velocity 5-inch/51-caliber guns was to protect the battleship from torpedo attacks by destroyers. Because these guns were not capable of defending the ship against air attacks, the *Arizona* also carried four 3-inch/50-caliber semiautomatic antiaircraft batteries and two underwater torpedo tubes. Her top speed was 21 knots, and she was powered by oil instead of coal. Her overall displacement was 32,500 tons of water.

USS *Arizona* was commissioned on October 17, 1916, and spent her first year with the Atlantic Fleet assigned to gunnery training. Three years later she joined the Pacific Fleet in California and spent the rest of her active service, aside from an eighteen-month overhaul in Virginia, in the Pacific.

In 1929 she was placed in a reduced-commission status and underwent major reconstruction to her defense against torpedo or underwater bomb damage. After her last overhaul in 1941, months before her fateful end, the *Arizona* had accommodations for 2,037 officers and enlisted men.

In postwar years, the nation turned its attention to the building of memorials to commemorate the events of the war years and to pay tribute to those who lost their lives. Hawai'i proposed a plan to build an *Arizona* memorial and the Pacific War Memorial Commission was formed. Funds came from the legislature and from donations solicited through the television program "This Is Your Life" hosted by Ralph Edwards; over $64,000 was raised in a sold-out Elvis Presley concert at Bloch Arena in Pearl Harbor.

The Navy stipulated that the new memorial be a type of bridge, with no part touching any of the sunken ship, and able to accommodate two hundred people. Architect Alfred Preis first submitted a design for a structure that allowed visitors to view the underwater remains of the *Arizona* from a subsurface structure. The Navy flatly rejected his proposal. Preis' second concept was approved, and the white marble bridge-like structure was erected above the *Arizona* and dedicated on Memorial Day 1962. The 184-foot Arizona Memorial has a shrine room at the back, with names etched in the marble wall of the 1,177 crewmen who lost their lives. Approximately a million-and-a-half people tour the site each year.

This bridge-like structure stands as a memorial to the sunken battleship below it—USS *Arizona*. One of the several targets of the Japanese attack on Pearl Harbor on December 7, 1941, the *Arizona* sank rapidly with most of her crew aboard. The memorial displays the names of 1,177 crewmen who lost their lives. Tours of the memorial are held daily.

• *See Historical Notes (page 127) for more information.*

USS Bowfin

On December 15, 1941, Congress ordered twenty-three new submarines, all of which were given fish names. USS *Bowfin* (SS 287) was one of them. She was also one of 288 U.S. submarines to see combat duty during World War II, operating in the Pacific theater, where naval and amphibious warfare was prominent.

Fleet-type submarines were actually "submersible ships," rather than submarines. They had long, narrow hulls with sharp bows, high reserve buoyancy and freeboard, large decks with guns mounted topside, and spacious bridges. Their speed on the surface was reduced by over 50 percent when they operated submerged.

The *Bowfin* was built in the Portsmouth Navy Yard, on an island in the Piscataqua River between Maine and New Hampshire. After her commissioning at Portsmouth, the *Bowfin* began combat operations in the South China Sea. Her first patrol led her to the Mindanao Sea, where she made her first strike on a Japanese cargo ship, the *Kirishima Maru*, which fired upon her unsuccessfully as she submerged. The *Bowfin* sank a cargo ship and a tanker and set another tanker on fire. She then went on to sink two smaller vessels and a two-masted schooner, then carried out two secret missions, delivering supplies to Philippine guerrillas and taking aboard nine guerrilla fighters to be transported to Australia. Upon leaving the Mindanao Sea, the *Bowfin* closed upon a small diesel barge with a Japanese flag and opened fire, first with her 4-inch deck gun and then with her 20-mm guns. The barge sunk immediately. During her first patrol alone, she was credited with sinking 23,753 tons of enemy shipping. On her second patrol, she encountered schooners, large ships, steamers, tankers, transport ships, and cargo ships. She took some hits while sinking the *Tonan Maru*, a 9,866-ton tanker, and had to return to Australia for repairs. She arrived in Fremantle on December 9, 1943, after being under way for thirty-nine days and sailing 10,323 miles. After this patrol her commanding officer, Lt. Comdr. Walter Thomas Griffith, was awarded the Navy Cross and the *Bowfin* received the Presidential Unit Citation.

On her third patrol the *Bowfin* covered 7,949 miles and sank 4,408 tons of enemy shipping. Lieutenant Commmander Griffith was promoted to commander and awarded a gold star in lieu of his second Navy Cross. The fourth patrol brought the *Bowfin* into contact with a convoy of four enemy ships, causing her to dive to 350 feet, where she managed to withstand a thorough shaking by depth charges. At one point the crew could hear the sound of a chain being dragged across the submarine's hull by the enemy vessel above.

Throughout all her patrols, there were constant problems with the torpedoes, which all too frequently exploded prematurely or didn't explode at all. Despite these obstacles, in a career of nine patrols, the *Bowfin* achieved record performance, with a total of thirty-eight ships destroyed, amounting to 67,882 tons: sixteen large ships and twenty-two smaller craft, according to official military assessments. Her four World War II commanding officers believed the total was higher: thirty-four large and ten smaller vessels sunk, for a total of 178,946 tons, and five large and two smaller vessels damaged. At the start of her tenth patrol, en route from Pearl Harbor to Guam, the Japanese surrendered and the *Bowfin* returned home to Pearl Harbor. She left Hawai'i in 1945 for the east coast and served with the Atlantic Fleet until January 1947, when she was placed in reserve.

After World War II, the *Bowfin* was a member of the mothball fleet in New London, Connecticut, for over four years after her decommissioning. But, as the Navy

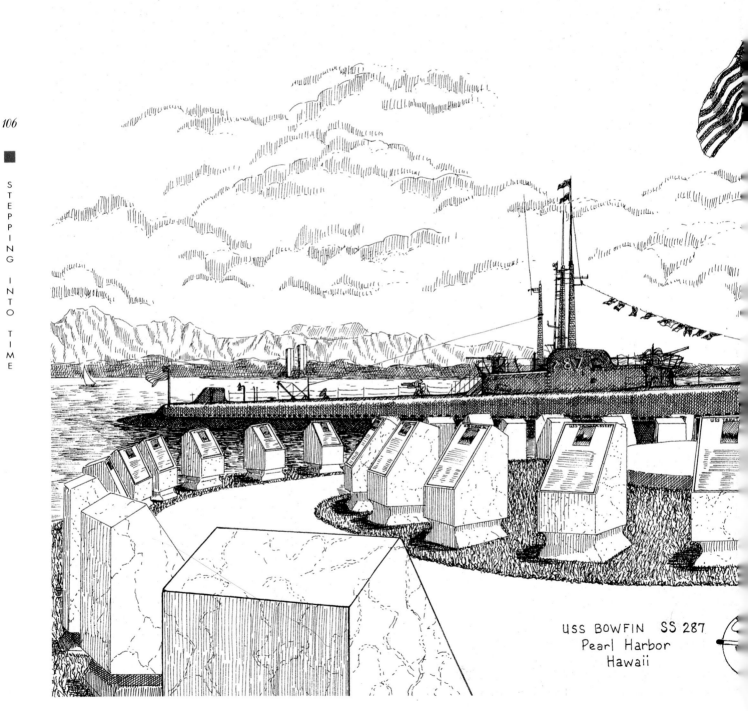

began to rebuild after the Korean conflict, she was recommissioned on July 27, 1951, and spent the next two years on training duty, operating out of San Diego as a unit of Submarine Squadron 3 and, later, Squadron 5. In 1952 she went through an overhaul at Mare Island Navy Yard in Vallejo, California, and her last years were spent as a Naval Reserve training vessel in Seattle, where she was moored to the training pier with her tubes and vents sealed. She was struck from the Navy list in December 1971 and towed to the Inactive Ships Maintenance Facility at Pearl Harbor, where she remained until 1979. Although most boats at this stage would have been sold for scrap, the *Bowfin* escaped fate again and was transferred to the Pacific Fleet Submarine Memorial Association, which proceeded to restore her for exhibition as a submarine memorial located in Bowfin Park, near the Arizona Memorial, at Pearl Harbor.

A short distance from Arizona Memorial Park at Pearl Harbor is Bowfin Park and the submarine memorial, USS *Bowfin*. This World War II fleet-type submarine, one of the few remaining submarine memorials on exhibit today, is open daily for tours. Bowfin Park also features a museum, gift shop, monuments to other World War II submarines, and exhibits of artillery and torpedoes.

• *See Historical Notes (page 128) for more information.*

War Memorial Natatorium

The idea for a World War I memorial in Honolulu originated years before the actual plans were enacted. As early as 1918, the Daughters and Sons of the Hawaiian Warriors proposed the creation of a memorial made of Hawaiian lava with polished sides, on which would be carved the names of all the islanders who gave up their lives in their country's cause during the Great War.

In 1921 the territorial legislature authorized the issuance of bonds to raise $250,000 for the construction of a memorial to the men and women of Hawai'i who served in World War I. The legislature further provided for the appointment of the Territorial War Memorial Commission to decide the form the memorial was to take. The legislature stipulated that a swimming pool of at least 100 meters, length be included and that a competition be held to determine the most appropriate design. The competition was held under the general rules of the American Institute of Architects. Three architects, Bernard Maybeck of San Francisco, Ellis F. Lawrence of Portland, and W. R. B. Willcox of Seattle, were selected to judge the competition. Louis P. Hobart of San Francisco won the first prize, but his plan had to be modified twice because of budgetary restrictions. It was not until 1927 that T. L. Cliff started construction.

The War Memorial Natatorium, constructed of reinforced concrete, contains an open-air, 100-meter by

The entryway to the War Memorial Natatorium depicted here opens to the first "live" war memorial in the United States—built to serve the living while commemorating the dead. It is located at the Diamond Head end of Waikiki, *makai* of Kapi'olani Park. Presently closed to the public, this historic site was once frequented by world-renowned swimmers and was a favorite recreational center for thousands.

• *See Historical Notes (page 128) for more information*

50-foot swimming pool fed by ocean water through a series of coffered locks. It uses the natural ocean bottom of sand and coral. The pool is surrounded on four sides by a 20-foot-wide deck, which is enclosed on the ocean sides by a 3-foot wall. On the *mauka* side, concrete bleachers rise thirteen levels and provide seating for approximately 2,500 people.

Perhaps the most impressive feature is the Beaux Arts style main entry, with its triumphal arch flanked by two lesser round arches. A pair of Ionic pilasters support the arch's entablature, which is inscribed with the words "The War Memorial." Arising from the entablature is a sculpture consisting of a garlanded base, with an American eagle perched at each corner and the Hawaiian motto and seal in the center. Flanking the triumphal arch and above the two lower arches is a medallion with floral patterns and a woman's face in the center in relief. The ocean and mountain sides of the entry are similar.

The Natatorium was completed in the summer of 1927. The opening ceremony held on August 24 was the major social event of the year. Duke Kahanamoku, Hawai'i's most famous swimmer, surfer, and Olympic gold medalist, traveled from Los Angeles to open the pool on his birthday, and he made the first swim, emerging at the end of the 100 meters amid thunderous ovation.

The Natatorium immediately became a social-recreational center for local people, and many international and national swimming meets were held there. In its first five years it was used by 700,000 people—for training scuba divers, lifeguards, policemen, and firemen in rescue work, for scouts tests, and just for swimming. It played a significant role in the "pan-Pacific" philosophy of the time, at least in athletics.

After the Pearl Harbor attack, the Natatorium was taken over by the U.S. Army and used for training exercises until 1943. Its physical condition continued to deteriorate and its neglect began to draw public attention. In 1949 it was rehabilitated by the territorial government at a cost of $82,000 and subsequently turned over to the City and County of Honolulu on July 1. Sporadic and inefficient maintenance continues to plague the Memorial Natatorium, which is presently in need of extensive rehabilitation. Although officially closed, the pool still attracts its share of visitors and remains a "living" memorial.

The War Memorial Natatorium is the only true memorial constructed by Hawai'i to its World War I casualties. The Punchbowl Cemetery of the Pacific and the Arizona Memorial are federal monuments relating to World War II and later years.

Washington Place

A COLONIAL MANSION STANDS IN DOWNtown Honolulu, obscured from busy Beretania Street by a wrought iron fence and an expansive, well-landscaped yard and surrounding gardens. It was built in 1846 by sea captain John Dominis, who had moved to Hawai'i from Schenectady, New York, with his wife and son in 1840. After Captain Dominis set sail for China and was never heard from again, a bereaved Mrs. Dominis sought to support herself and her son, John Owen, by renting a suite in their home to the American commissioner, Anthony TenEyck, and his son. Perhaps it was the mansion's resemblance to Mount Vernon that inspired Commissioner TenEyck, with Mrs. Dominis' approval, to name the home Washington Place after the first president of the United States. The name was made official on February 22, 1848.

Young John Owen Dominis worked for G. B. Post Company of San Francisco during the Gold Rush, and later for R. Cody & Company, ship chandlers in Honolulu. Dominis was appointed secretary and chamberlain to King Kamehameha IV and, upon the death of Kekuanao'a in 1868, Kamehameha V appointed him governor of Hawai'i. Shortly before he became governor, he brought his bride, the High Chiefess Lydia K. P. Kapa'akea, home to Washington Place. Soon thereafter, young Mrs. Dominis was given the title of Princess Lili'uokalani and named heir to the throne by her brother, King Kalakaua. She acted as regent of the kingdom during Kalakaua's world tour and, when he died in 1891, she assumed the throne. John received the title of prince consort. Scarcely eight months later he died, leaving Washington Place to his wife.

In her early years of residence at Washington Place, Lydia Dominis described it as:

"a large, square, white house with pillars and porticos on all sides, really a palatial dwelling, as comfortable in its appointments as it is inviting in its aspect; its front is distant from the street, far enough to avoid the dust and noise. Trees shade its walls from the heat of noonday; its ample gardens are filled with the choicest flowers and shrubs; it is, in fact, just what it appears, a choice tropical retreat in the midst of the chief city of the Hawaiian Islands."

Washington Place gained a reputation as a special rendezvous of royalty, visiting dignitaries, and heads of state. At the suggestion of Prince Jonah Kuhio Kalaniana'ole, an heir to the estate of Queen Lili'uokalani, the Territory acquired Washington Place as the executive mansion. It became the official home of the governor of Hawai'i on April 21, 1922, and has often been called the "Little White House of the Pacific."

Washington Place holds many memories of a romantic period when daring sea captains came to Hawai'i and made it their home, and when foreign influence eroded the foundation of a once powerful monarchy. As a poignant reminder of the beloved Queen Lili'uokalani, a plaque designed by sculptor Kate Kelly and inlaid in a lava boulder on the grounds of Washington Place carries the words to her song, "Aloha 'Oe."

• *See Historical Notes (page 129) for more information*

Washington Place, Honolulu, Hawaii

Washington Place, home of Hawai'i's governors, was once occupied by the last Hawaiian monarch, Queen Lili'uokalani. Situated next to St. Andrew's Cathedral on Beretania Street, the charming home dates to 1846, when it was built by sea captain John Dominis.

Historical Notes

ALEXANDER & BALDWIN BUILDING

Thomas Alexander (1836–1904) was born on Kaua'i and educated at Punahou School and at Williams College in Massachusetts. He began his career as a plantation manager for the Waihee Plantation on Maui. Henry P. Baldwin (1842–1911), also educated at Punahou and Williams College, became head of the plantation and later Alexander's brother-in-law and business partner. In 1869 the two formed the Haiku Sugar Company, which invested in an elaborate 17-mile irrigation system on Kaua'i known today as the Hamakua–Haiku Ditch. The company expanded to form a new firm in San Francisco under the name Alexander & Baldwin. In 1897 a branch was established in Honolulu, and three years later the firm was incorporated as Alexander & Baldwin, Ltd., with Henry P. Baldwin as president. During its early years, Alexander & Baldwin operated a fleet of ships between Hawai'i and the mainland. This operation was later taken over by one of its subsidiaries—the Matson Navigation Company.

ALOHA TOWER

When Pearl Harbor was bombed by the Japanese in 1941, Aloha Tower was closed to all but officials. It became the center for planning military convoy operations in the Pacific theater, and many Pacific Island invasions were planned within its offices. Because of its strategic significance, it was covered with green and khaki camouflage paint during the war years; when it reopened in 1948, the paint was removed and the original color restored.

BERNICE PAUAHI BISHOP MUSEUM

Bishop Museum's original holdings were kapa, mats, calabashes, featherwork, ornaments, and relics of the royalty bequeathed to Bernice Pauahi Bishop. Later other collections included those contributed by J. S. Emerson, G. H. Dole, Eric Craig, and the estate of Queen Emma, the Garrett Collection of more than 9,000 species of shells, and the Mann and Brigham Collection of Hawaiian plants. Bishop Museum's first curator, William Tufts Brigham, wanted a museum "to instruct and delight." He succeeded in transforming "a cabinet of curiosities" into a permanent source of historical and archaeological relics and natural specimens from the entire Pacific Polynesian region. Among its treasures were feather capes from the Kamehamehas, feather leis and helmets, temple idols, carved calabashes, royal crowns, thrones, scepters, and swords. Some of these items have since been restored to 'Iolani Palace. Exhibits of Polynesian navigation, historic artifacts, and specimens of birds, plants, and marine life were later added to the museum.

C. BREWER BUILDING

The following excerpt from the Hawaiian periodi *The Friend*, of January 1867, tells about the entrepreneurial st of Captain James Hunnewell, whose trade and shippi business became the C. Brewer & Company:

I commenced my seafaring life in 1809. In October, 1816 I embarked in the brig Packet for San Francisco, expecting to touch at the Sandwich Islands for supplies early in 1817. Touched at Hawai'i, Maui, and O'ahu, procured our supplies, and proceeded to the coast, and up the Gulf of California to Loretto— thence round and up the coast. After sundry escapes from capture, returned to Honolulu and visited Kamehameha I. At Hawai'i, and after various negotiations, our brig was sold, and paid for in sandal-wood, which required several trips around O'ahu, the wood being nearly all shipped on ships at Honolulu. Our captain, A. Blanchard, embarked for China, leaving Mr. Dorr, my fellow officer, and myself, to remain and dispose of the balance of California cargo, and ship the proceeds (sandal-wood) to China. All trade was in barter, as there was no money in circulation among the natives. This detained us until September, 1818, when I embarked in the ship Osprey, and Mr. Dorr in the ship, Enterprise, to China, with our sandalwood. We were the only traders on shore at Honolulu that had any goods to sell. All our cash sales amounted to $104, and this was from an English captain and officers.

At first, C. Brewer & Company supplied barrels for the shipment of sugar and molasses. After acquiring a plantation on Maui, the company became agent for three Maui plantations, and by 1883 it handled 14 percent of Hawai'i's sugar trade. Brewer's nephew, Charles Brewer II, took over the business in 1858. In 1871 Peter Cushman Jones became a partner, and Charles Reed Bishop entered the firm in 1880. Jones took over as president in 1883 and served until 1899, when he was succeeded by Charles M. Cooke.

CATHEDRAL OF OUR LADY OF PEACE

The nineteenth-century Protestant missionaries strongly opposed the introduction of Catholicism to the islands. In their minds, acceptance of the Catholic faith was tantamount to treason, for Protestant Christianity was virtually the national religion. Furthermore, Catholics were considered idolaters who worshipped crosses and statues. The Protestant missionaries took measures to convince the ruling chiefs and Queen Ka'ahumanu to force the first Catholic priests to leave. When the priests refused, they were imprisoned along with their parishioners. The persecution of the Catholics lasted for two years, and on Christmas Eve in 1831 the priests finally departed aboard the small Hawaiian brig *Waverley*.

A second attempt to establish the Catholic mission was made in 1835, when Brother Murphy was sent by the newly appointed vicar apostolate, Bishop Rouchouze. Since Murphy was a British citizen, there was no objection to his arrival, and he was allowed to go about freely. He was followed by Father Arsene Walsh in 1836, an Irishman who inspired Bishop Rouchouze to send the exiled priests back to the islands, believing that they might fare better this time. But, when Fathers Bachelot and Short returned in 1837, they were ordered to return on the ship that brought them.

In 1839, French commander LaPlace sent a strong manifesto to the Hawaiian government, demanding that Catholic worship be permitted and that Catholics be allowed the same privileges granted to Protestants. Fearing repercussions from France, King Kamehameha III surrendered to pressure and granted freedom of religion for all people living in the Hawaiian kingdom. Catholic natives were released from prison and priests returned to the islands.

In this cathedral, on May 21, 1864, a young Belgian named Joseph de Veuster was ordained a priest. He chose as his priest's name Damien and requested to serve as a missionary at the leper colony on Molokai. There Father Damien built a church and lived the rest of his life serving the victims of leprosy.

DILLINGHAM TRANSPORTATION BUILDING

Walter F. Dillingham, son of Franklin Dillingham, was educated at Punahou School and Harvard University. He joined his father's rail company as a clerk and later became the first manager of the Hawaiian Dredging Company. His business interests were concerned mainly with developing harbors at Honolulu, Hilo, Kahului, and Ahukini. He is also credited with improving the Pearl Harbor naval base and building a huge drydock there.

The Dillingham name is also associated with the Ala Moana Shopping Center complex. La Pietra, Walter and his wife Louise's home on the slopes of Diamond Head, was for years a social gathering place for visiting celebrities and is today the home of the Hawai'i School for Girls.

'EWA PLANTATION

'Ewa Plantation's early history is intricately tied to three men: James Campbell, Benjamin Dillingham, and W. J. Lowrie. Campbell envisioned the idea, brought irrigation to the arid lands, and built the sugar cane milling plant at 'Ewa; Dillingham provided rail transportation with the establishment of the O'ahu Railway & Land Company; Lowrie was the plantation's first manager.

Dillingham envisioned a rail transportation system linking the outlying farm districts with the city and waterfront of Honolulu. He leased land from James Campbell at both Kahuku and Honouliuli for fifty years, beginning in 1889, for $50,000 per year. That same year he obtained a charter from the Hawai'i government for the O'ahu Railway and Land Company. Next, he endeavored to convince W. R. Castle of the potential value of cultivating sugar cane in 'Ewa. On January 29, 1890, the 'Ewa Plantation Company was chartered. Under the management of W. J. Lowrie, the rangeland was cleared, plowed, and cultivated, and buildings were erected to house the plantation workers, blacksmith, and farm equipment. It was probably one of the worst times to begin such an enterprise. There were problems of labor shortage, plummeting sugar prices, and political turmoil that eventually culminated in the 1893 overthrow of the monarchy. Nevertheless, 'Ewa Plantation forged ahead. By 1892 the plantation company was in debt for an estimated $1,200,000 with no hope of recovery. The stockholders were enraged and the banks were refusing credit. Both 'Ewa Plantation and its major source of revenue, the firm of Castle & Cooke, faced bankruptcy until the latter was able to obtain help from a San Francisco firm. The situation steadily improved and, by 1898, thirty-five wells were in service and the plantation reached an output of 19,583 tons of sugar—a record crop for the islands. The company had become debt-free at last.

When Lowrie retired, George F. Renton from Kohala Sugar Company took over as plantation manager. Renton's interests turned toward the development of technology to improve cultivation and milling and experimentation to develop sugar cane varieties. Through his efforts, camps for the workers were replaced with real homes and schools and support facilities were built, and the plantation community became a model for its time. The results of his experimentation brought new strains of sugar cane, and a popular variety called H-109 revolutionized the industry. It was a disease-resistant, high-sucrose variety that was soon used throughout Hawai'i.

When Renton retired in 1920, his son, George, Jr., took over management. He upgraded and modernized the plantation and brought it to new standards of comfort and sanitation for the employees. Homes were remodeled, electricity installed, and new coral roads and a health facility added. Renton also established a department of agricultural research and control. By 1922 the mill village was completed, with a total of 642 dwellings equipped with electricity and other modern conveniences. That year also marked a record crop for the plantation.

FOSTER BOTANIC GARDEN

Dr. William Hillebrand was instrumental in bringing new varieties of plants to the islands. In his first report to the Society on Floriculture, he pointed out the variety of plants and shrubs growing in remote areas of the islands which could be transplanted in populated areas where they could be enjoyed by everyone. He urged islanders to procure seeds from other tropical countries and praised the efforts of H. Bridge, purser of the U.S. sloop-of-war *Portsmouth*, and Dr. Brinckerhoff, who had brought a variety of seeds to Honolulu.

Although Hillebrand was primarily concerned with horticulture, he may be considered Hawai'i's first environmentalist: he stressed the idea of sanitation in Honolulu. In one of his discourses on flowers, he interjected, "May the city fathers . . . keep their eyes and noses open . . . to speak of fragrance and stench in one breath, for I mean to expel the latter by the former." He suggested that the citizens of Honolulu line the principal streets in and near Honolulu with fine shade trees, thus checking the high winds crossing the Pali and also serving as an effective barrier to the spread of fire. He advocated the distribution of lots to landless natives so that they could establish homesteads. In view of the dwindling Hawaiian population, which was being decimated by foreign diseases, he also urged the founding of adequate hospitals. Hillebrand and Dr. J. Mott-Smith opened a temporary hospital and dispensary near the foot of Fort Street, and later Hillebrand became head physician at the new Queen's Hospital.

Through Hillebrand's efforts and the support of the Royal Hawaiian Agricultural Society, one part of the Foster Garden story evolved. The other part involves Mary Elizabeth Mikahala Robinson Foster, eldest daughter of James Robinson, a well-to-do Honolulu shipbuilder, and his part-Hawaiian wife, Rebecca Prever. Mary married Thomas R. Foster, a Canadian born in Pictou, Nova Scotia, also active in the shipping business, and they settled in their Nu'uanu home, next-door to Hillebrand. When the latter left Hawai'i in 1871 to travel to Europe and eventually reside in Madeira, the Fosters bought Hillebrand's property, with its beautifully planted garden, and built a new home on it. Captain Foster erected a five-story tower next to their mansion from which he could keep an eye on his ships in Honolulu Harbor. Mary Foster spent much of her time working in the gardens, where flourished not only trees and plants, but monkeys and a giant tortoise on which the Robinson nieces and nephews used to ride. She added many new plants and trees, but her favorite was a bo tree, said to be a cutting of the original bo tree under which Buddha was enlightened. Mary kept the grounds in good order until her husband passed away and she left to live with her sister. The gardens soon fell into neglect, and a jungle eventually replaced the once scrupulously tended grounds. In 1918, after Mary had moved back to the property, she enlisted the help of Dr. Harold L. Lyon, an employee of the Hawai'i Sugar Planters' Association, and from his labors the present Foster Botanic Garden began to grow.

HAWAIIAN ELECTRIC COMPANY BUILDING

Electricity had first been demonstrated in Hawai'i in 1886, at a time in history when steamer fare from Honolulu to San Francisco was $25 one-way and horse-drawn tram cars ran between Waikiki and the bustling Honolulu waterfront. Five years earlier, in 1881, King Kalakaua had the opportunity to meet with Thomas Edison, who joked with the king about harnessing the power of Kilauea volcano to provide electrical power for all of America. The first demonstration of the new technology was a spectacular social event that brought crowds to the grounds of 'Iolani Palace to witness the illumination of arc lamps hung at the palace, its Richards Street gate, Ali'iolani Hale, and King Street. By 1888, King Kalakaua's government was operating a hydroelectric plant on Nu'uanu Stream which sporadically provided electric power, depending on the amount of rainfall in Nu'uanu Valley.

HONOLULU ACADEMY OF ARTS

Anna Charlotte Rice, the founder of the Honolulu Academy of Arts, was born to missionary parents, William Harrison Rice and Mary Sophia Hyde, on September 5, 1853. She developed a deep love of music, literature, and art early in life.

Anna was educated at Punahou School and at Mills College in Oakland, California. In 1874 she married Charles Montague Cooke, Sr., also a child of a missionary family. They settled in Honolulu and built their first home on the outskirts of town, opposite Thomas Square on Beretania Street. The Cookes had six sons and two daughters. As her children grew, Anna Cooke broadened her art collection and displayed it in her home for the enjoyment of friends and invited guests. Starting with mostly Western art, she gradually acquired some Asian pieces and her interest in it grew. Anna's compelling desire was to instill in island children an understanding of their cultural heritage and to introduce them to the art of other cultures. After her husband died in 1909, her sole interest was to share the fulfillment of the art experience.

By 1920 the Cooke residence could no longer hold all the art treasures Anna had collected, and she began planning a small museum with Mrs. Isaac Cox, an art and drama teacher. Together, they researched the art objects and catalogued them for a time when they could be publicly displayed. In 1922 the Territory of Hawai'i issued a charter of incorporation to Anna's museum, the Honolulu Museum of Art.

'IOLANI PALACE

There was strong opposition to the building of 'Iolani Palace, particularly by non-Hawaiian businessmen and government officials who resented the king's extravagant spending and his desire to rule the kingdom rather than simply reign in a foreign-controlled country. King Kalakaua's popularity was a deterrent to those seeking government control, and his determination to preserve the ancient traditions of Hawai'i was viewed by some as heathen and irreligious.

The old, coral block-and-wood palace he planned to replace had been built in 1844 in the area known as Pohukaina, "The Place of Ordered Calm." The building was called Hanailoia. In olden times this had been the site of the *heiau* of Kaahaimauli, a land sacred to the Hawaiians. When the capital of the kingdom was moved from Lahaina, Maui, to Honolulu in 1845, King Kamehameha III took possession of this residence and made it his royal home. He changed its name to Hale Ali'i, "House of the Chief." The name 'Iolani Palace, "Palace of the Bird of Heaven," was adopted later.

Kalakaua decided to be the first king of Hawai'i to be officially crowned. While in England, he ordered two golden crowns set with precious jewels. When he returned from his world tour, he proceeded with plans for a lavish grand opening and coronation ceremony. A special pavilion surrounded by an amphitheater was constructed adjacent to the palace, and there Kalakaua crowned himself king and his wife queen on February 12, 1883. A celebration lasted two weeks and included a coronation ball, fireworks display, *hula* dancing, a regatta, horse racing, and endless feasting and entertainment. When the celebration was over, the coronation pavilion was moved to a new location on the palace grounds and the amphitheater was dismantled.

The king and queen hosted many a festive evening at 'Iolani Palace. Making their grand entrance down the koa staircase, the royal couple would lead the entourage into the throne room, where music and dancing continued until the king bade his guests farewell. An more intimate gatherings, there was dancing in the ancient *hula* tradition. Kalakaua himself would instruct and explain the mystic meaning of the dances. When word got around that the *hula* was being taught and practiced at the palace, outraged moralists cried out against the perpetuation of heathen chants and licentious practices.

King Kalakaua formed a society in which *kahuna kuauhau*, historians, transcribed the ancient Hawaiian records that had been passed along from generation to generation through chants and *meles*. In this society, the Kumulipo, a Polynesian chant of creation, was chanted by venerable *kahunas* and transcribed by scholars proficient in the ancient vocabulary. Kalakaua's interest in esoteric lore led to a revival of Ka Hale Naua, "The Temple of Wisdom," devoted to the research of human origins.

The golden era of King Kalakaua ended when he died on January 20, 1891. Final services were held in the throne room, and over his casket the golden cloak of King Kamehameha was laid. Next to him were placed his crown, scepter, and crown jewels, and on either side ten majestic *kahili* were held by bearers. The *Daily Bulletin* recalled: "He was a true and loyal king. . . . he died as he had lived, with an eye single to national advancement. . . . Hawai'i is filled with mourning."

Queen Lili'uokalani, Kalakaua's sister, succeeded him on the throne until 1893, when the monarchy was overthrown by the then-powerful oppositional faction that set up a provisional government. In 1895, after an unsuccessful attempt to restore her to the throne, Queen Lili'uokalani was arrested and imprisoned for nine months in an upper-story room of 'Iolani Palace. The name of the building was changed to "Executive Building," and it was drastically altered to serve government purposes. Its furnishings, fixtures, chandeliers, and artwork were auctioned off, along with portraits and possessions of the royal family. Carpets were cut into pieces and sold. Most of its rooms were sectioned into offices, and the exquisitely designed interior was concealed under coats of institutional green paint.

'IOLANI BARRACKS

In September 1873 there was a mutiny of the fifty-man Royal Guard at 'Iolani Barracks during the reign of King Lunalilo. Some people believed that David Kalakaua had a hand in the mutiny, intending to humiliate King Lunalilo's government, which he considered too friendly with "foreigners" and too pro-Western. The six-day mutiny was the direct result of a bitter dislike of the overly strict Royal Guard drillmaster, Capt. Joseph Jajczay, a martinet trained in the Austrian army. There was also resentment over some acts of the adjutant general, Col. Charles H. Judd. The Royal Guard barricaded themselves inside the barracks and refused to give up their arms. The only ones allowed to enter were the Royal Hawaiian Band. To end the mutiny, the king, who was ailing at his royal beach house in Waikiki, had to disband the Guard. Five months later, King Lunalilo died and David Kalakaua assumed the throne and reinstated the Royal Guard, which he named the King's Guard.

KAKA'AKO PUMPING STATION

The need for an adequate sewage system in Honolulu was heightened by the outbreak of bubonic plague in Chinatown in 1898. Several residents had contracted the disease, which was spread mainly by fleas carried by rats that proliferated in the unsanitary conditions of the city. The board of health's method of controlling the plague was to evacuate an area in which victims had died and burn it to the ground. The tragic fire of January 20, 1899, which claimed most of Chinatown and left seven thousand people homeless, started when one of the disease-controlling fires raged out of control.

Honolulu's sewage situation had been a subject of controversy long before the 1900s. The monarchy studied the problem, the 1880 legislature authorized a sewage system that never materialized, and the Department of the Interior again studied the problem in 1890 with no action taken. The thrust behind a movement to correct the situation came after the overthrow of the monarchy in 1893, when Sanford Dole and his Committee of Safety were seeking annexation by the United States. They reasoned that a permanent public works department with a well-designed sewerage system would contribute to Hawai'i's acceptance as a U.S. territory. Two more epidemics, this time of cholera, were to sweep the city before the government finally moved to establish the long overdue sewerage system.

KAMEHAMEHA V POST OFFICE

H. M. Whitney, Hawaii's first postmaster, resigned in 1856 to form the *Pacific Commercial Advertiser*, now *The Honolulu Advertiser*. Whitney later became distinguished as the "Father of the Hawaiian Post Office."

In the early days of the postal service, private mail was delivered in leased boxes and general delivery mail was dumped in a pile on the floor which everyone dug through to find their own. The captain of the ship delivering mail was paid two cents for every letter and one cent for each newspaper delivered. When Whitney gave up his post office business in 1856, the position of postmaster went to J. Jackson, who was later succeeded by A. K. Clarke. David Kalakaua held the position briefly in 1865, followed by A. P. Brickwood.

A story was told about a man who rode his horse over the new concrete sidewalk at the post office and was summarily reprimanded and charged a fine of two dollars "by way of warning to all future horsemen." On another occasion, a curious, visiting passer-by spied two cannons lying in front of the post office and inquired as to their purpose. He was informed that one cannon would be fired ten minutes before the opening of the general delivery and the other ten minutes before the closing of the office. The befuddled stranger walked away musing over the queer customs of Hawai'i, not knowing that the cannons were used as hitching posts. They were believed to have come from the old Honolulu fort.

As early as 1945, the old post office building was described "as rich in historical associations as it has been in rats, termites and cockroaches." The condemned building was subsequently remodeled and returned to use by the traffic courts and, in 1966, with its upper floor condemned, it continued to serve as temporary quarters for the district courts. It also housed state welfare and driver education offices. The Kamehameha V Post Office was placed on the National Register of Historic Places on May 5, 1972.

THE KAMEHAMEHA SCHOOLS

A board of five trustees named in Princess Bernice's will was directed to carry out her wishes for the building of the Kamehameha Schools. They agreed upon four basic principles: that the school be in Honolulu, that it be Christian, that its students learn to read and write English clearly, and that it be a practical school teaching children to lead useful lives.

The original Kamehameha School for Boys was built in Palama, on the site of Bishop Museum. It opened on November 4, 1887, with thirty-seven students. King Kalakaua and Princess Lili'uokalani were among the members of the royal family attending the opening exercises. The campus consisted of five frame buildings: two dormitories, a dining hall, the principal's house, and a workshop. The first principal was Rev. William Brewster Oleson. After the School for Boys opened, Charles Reed Bishop, realizing he had overlooked younger, homeless boys, established the Preparatory Department, which opened on October 29, 1888, with Miss Carrie A. Reamer as the first principal. The third project was the building of the School for Girls *makai* of the School for Boys across King Street. The trustees decided that it would be "a school to teach employment, such as cooking, dressmaking, care of house, etc. etc. for older girls." The School for Girls opened on December 19, 1894, with thirty-five girls enrolled, under the direction of Principal Ida May Pope.

KAWAIAHA'O CHURCH

According to Rev. Hiram Bingham, each of the thatched-roof churches that preceded Kawaiaha'o Church on the same site was "a long haystack without and a cage in a haymow within." The present church was dedicated on July 21, 1842, without Reverend Bingham present. After twenty-one years of service in the islands, he had returned to New England in 1840 because of his wife's failing health. He never returned to Hawai'i.

Through the influence of the missionaries, churchgoing had become a sort of holiday for the *ali'i*. They arrived at Kawaiaha'o Church dressed in every piece of finery imaginable. Queen Ka'ahumanu, her colossal figure draped in yards of yellow satin and purple silk, would be trailed by chiefs and chiefesses of lesser rank wearing anything from British uniforms to gowns of silk or calico. Heads were adorned with French chapeaux topped with ostrich plumes. Feet were clad in leather boots, or even high-laced woodsman's boots. Some wore cutaway coats trimmed in gold braid above bare brown buttocks, or costumes of loincloth and necktie only.

Kawaiaha'o Church was the scene of many historic events. As the state church and house of worship for the royalty for twenty years, it was the scene of weddings, funerals, thanksgiving ceremonies, and public meetings.

Kamehameha III broke an old tradition by worshipping in a ground-floor pew of Kawaiaha'o Church while others sat above him in the galleries. He wanted a place near a window and the pulpit, he said, and he didn't care who sat in the galleries as long as they didn't break through and fall on his head.

It was at Kawaiaha'o Church that the first Hawaiian, Rev. James Kekela, was ordained minister in 1849. Here, Prince Alexander Liholiho gave his accession speech when he became King Kamehameha IV. Here, he took for his bride the lovely Miss Emma Rooke in 1856. William Charles Lunalilo, known as "Prince Bill," walked to Kawaiaha'o Church amid cheering crowds to accept the throne to which he had been elected. And here, the bodies of Kamehameha III, Queen Emma, Princess Ka'iulani, Queen Lili'uokalani, and many other notables of the kingdom lie in state.

LA PIETRA

For four decades La Pietra was the social center and showplace of the Dillinghams, who hosted many parties and entertained well-known guests such as Franklin D. Roosevelt, Herbert Hoover, the Prince of Wales, the Prince and Princess of Sweden, and the King of Thailand. Actors and actresses, musicians, composers, dignitaries, and top business and military leaders came to La Pietra to enjoy the finest of island hospitality. Virtually anyone well known at the time who came to Hawai'i was invited to La Pietra. During World War II, La Pietra became an official military headquarters, where military and political leaders met to discuss the progress of the war. Generals Marshall, MacArthur, and Emmons and Admirals Halsey, Nimitz, and Andrews all stayed at La Pietra during the war years.

With its purchase and renovation by the Hawai'i School for Girls, La Pietra made a remarkable transformation from a private home to a private school. It is certainly one of the most beautiful schools in Hawai'i and has continued its reputation as a center for art and theatrical and musical performances. Since its first restoration, the school has expanded to include a gymnasium, a library-laboratory building, and the Lorraine Cooke Hall, all designed by architect Leo Wou.

LINEKONA SCHOOL

Public education in Hawai'i can be traced back to 1831, when the O'ahu Charity School was established to teach English to half-whites. For many years it was the only English-language school in Hawai'i and one of six such schools west of the Rocky Mountains. O'ahu Charity School was renamed the Town Free School in 1851. It later became the Mililani Girls School after the board of education decided to separate school children by gender. The boys were sent to either the Royal School or the new, private Fort Street School.

Fort Street School was known as a select school, with classes conducted in English and a tuition charge of fifty cents a week. The first public high school came into being in 1895, when the Fort Street School became public and divided into Ka'iulani Elementary and Honolulu High School. Before this, secondary schooling had been left largely to such privately controlled and supported institutions as Punahou, 'Iolani, and St. Louis.

Honolulu High School met at the former palace of Princess Ruth Ke'elikolani on Queen Emma Street until 1908, when a permanent building was constructed on the Maertens property facing Thomas Square—the present site of Linekona School. It was originally dedicated as McKinley High School in honor of the U.S. president who held office during the annexation of Hawai'i as a territory. When the high school moved to its present campus on King Street, the vacated building became Linekona Elementary School until 1956.

LUNALILO MAUSOLEUM

During Lunalilo's reign, there was a mutiny by members of the Royal Guard. Through the king's appeal the revolt was settled, but the guard was subsequently disbanded, leaving the kingdom without a military force. The humiliation of the event, together with a history of ill health, led to King Lunalilo's untimely death at age forty-two. On February 3, 1874, his short reign of thirteen months came to an end.

The king was remembered by his people as a bright and gentle man, affectionately called *ke ali'i lokomaika'i*, "the kind chief." Though some found him to be vulgar and a drunk, others, like Mark Twain, found him "charming and cultivated." Lunalilo never married and left his personal wealth to the founding of a home for the aged. In his words, it was for the "poor, destitute, and infirm people of Hawaiian blood or extraction, giving preference to old people." Lunalilo Home at Koko Head is his tribute to all Hawaiians.

While King Lunalilo's tomb was being built, his remains were temporarily kept in the Royal Mausoleum. On November 23, 1875, Lunalilo's casket was transferred to the new tomb according to custom, shortly after midnight. The same procedure had been followed in 1865, when the remains of the royalty were transferred from the old mausoleum near ʻIolani Palace to the new location in Nuʻuanu. A story is told of how Lunalilo's father requested that his son be given the customary twenty-one gun salute upon his interment. King Kalakaua denied the request on the grounds that Lunalilo had already received a royal salute when he was interred at the Royal Mausoleum. Because he was no longer king, the show of respect was unnecessary. Nevertheless, as the hearse carrying the casket entered the Kawaiahaʻo churchyard, onlookers claimed to hear a loud peal of thunder, followed by twenty more distinct claps of thunder. The Hawaiians declared it "a heavenly royal salute to their beloved King Lunalilo." Newspaper headlines the next day read, "Kalakaua Rules; The Gods Overrule."

MCKINLEY HIGH SCHOOL

McKinley High School served more than half the high school students in Hawaiʻi through the 1920s. It was intended to provide a level of education to the general public previously attainable only at private schools. The school's population was largely non-Caucasian and racially integrated due to the efforts of its first principal, Miles E. Carey, who served from 1924 to 1948. Carey contributed significantly to the educational system by developing a core curriculum of English and social studies and by encouraging democratic participation by students in all learning activities and local government processes. Carey's program was thought to be liberal and considered by some to threaten the existing structure of the government.

Many McKinley graduates went on to become leaders in the community, establishing themselves in prestigious careers in government and politics. Among these were U.S. senators Hiram L. Fong and Daniel K. Inouye, Governor George Ariyoshi, and Hawaiʻi Chief Justice Wilfred Tsukiyama.

MISSION HOUSES

The Protestant missionaries who came to Hawaiʻi were directed to bring the Gospel to the heathen natives, teach them the rudimentary skills of reading and writing, and help them to establish comfortable dwellings. When the first company arrived in 1820, King Kamehameha II allowed them to settle on a small parcel of land on the outskirts of the village of Honolulu.

The first house they erected, called the frame house, was occupied by four families: the Daniel Chamberlains and their six children, the Loomises, the Thurstons, and the Binghams. Lucy Thurston, one of the first residents, expressed her relief at having a structure with doors that could be shut against the ever-present curious natives. The wallpaper, stuck on with poi, was the gift of a sea captain. The kitchen was added later, so until 1823 the missionary wives cooked meals outside, while natives stared in amazement at the curious scene. In the native culture, men did all the cooking: the sight of these thin, white women with their bonnets and long necks, stirring large kettles over a fire, was a source of wonder. The Hawaiians devised a name for them—the "long-necks."

In one period there were twelve adults and twelve children living under one roof, with as many as fifty people gathering for meals three times daily at the "long table" in the cellar. Meals consisted of salt pork, salt beef, and moldy ship's biscuits, which were made more palatable with a spoonful of molasses. Besides nurturing their own families, missionary women cared for orphaned native babies and sick and dying sailors. Each year, missionary families from all the islands gathered at the mission station in Honolulu, sharing the already crowded living quarters and meager fare with their brethren. To maintain good relations with the royal family, the missionary wives sewed dresses for the queen and ruffled shirts for the king. Missionary husbands wrote sermons and reports and struggled to learn the Hawaiian language in order to translate the Scriptures into Hawaiian.

The missionaries realized that the success of their work would depend on their ability to teach the Gospel in the Hawaiian language. They brought with them a second-hand Ramage press and a young printer by the name of Elisha Loomis. Their small printing house was completed in December 1823. This was the start of an educational revolution that would soon give Hawai'i one of the highest rates of literacy in the world. By 1825 the volume of printing had increased so enormously that Loomis wrote to the Mission Board that an addition to the printing house was necessary.

Levi Chamberlain had the responsibilities of securing and storing food, clothing, furniture, and other supplies needed by all the island stations and of distributing them as required. The parlor of his home served as a council room, a study, and an assembly room for family devotions and for the reception of royalty and other guests.

The missionaries introduced a new moral standard, reoriented the agricultural and technical skills of the natives, advised the royalty in political and economic matters, and taught most of the population to read and write. When their mission was deemed accomplished, the American Board of Commissioners for Foreign Missions disbanded the Hawai'i mission in June 1863. The mission families continued to live and pursue their livelihoods in the islands, some going into business and commerce.

MOANA HOTEL

In 1896, when Hawai'i was still an independent republic, its popularity as a vacation spot and gathering place for the elite was beginning to grow. To accommodate this new wave of tourism, plans were made for a grand hotel to be called the Moana, meaning "the broad expanse of the ocean." When the hotel was completed in 1901, the *Hawaiian Annual* described it this way: "From the beautiful grounds near the street the building rises majestically to its great height; its perfect delineation, graceful carving and elegant finish accentuated by the bright sunbeams, formed a picture not soon to be forgotten."

The Moana has undergone many changes since its early days. Its most recent renovation received many awards for historic preservation, including the National Preservation Honor Award in 1989, the Hawai'i Renaissance Award, the National Renaissance Merit Award, the American Society of Interior Designers Hawai'i Chapter Award of Merit, the Historic Hawai'i Foundation 1989 Preservation Award, and the President's Historic Preservation Award. The Governor's Commission on Persons with Disabilities presented the hotel the Exemplary Architecture Design Award, recognizing its efforts in creating a barrier-free environment. The Moana was the first hotel to receive this award.

Upon its reopening in March 1989, a press release issued by the Advisory Council on Historic Preservation in Washington, D.C., stated that the Moana project "restored the historic integrity of the illustrious hotel, bringing its grand public spaces back to life while simultaneously incorporating modern appointments and life-safety equipment. The newly-refurbished hotel now evokes its own and its island setting's proud heritage as it eloquently conveys its message to thousands of visitors: New is not necessarily better."

O'AHU RAILWAY AND LAND COMPANY

Benjamin Dillingham's promise to give his friends a ride on his new railway on his forty-fifth birthday was kept, though not exactly the way he intended. The "engineer" of Kauila No. 6 was an investor in the OR&L and untrained in steam locomotive operation. When all the invited guests were aboard, dressed in their finest attire for the occasion, the engineer blew the whistle, pulled back on the throttle, and engulfed everyone in a cloud of greasy black smoke. As they reached the Palama rice fields, he opened the throttle to increase speed and this time a shower of soot drenched the guests. Dillingham's birthday and the maiden trip of the Kauila No. 6 were not soon to be forgotten.

In 1971 a group of volunteers formed the Hawaiian Railway Society, a nonprofit educational organization committed to establishing an operating railway museum to preserve this aspect of plantation life. The group has been active in restoring the remaining engines, cars, and rails of the OR&L. Through their efforts the section of track between 'Ewa and

Nanakuli is now listed on both the State and National Registers of Historic Places. On November 16, 1989, the Oʻahu Railway and Land Company celebrated its one-hundredth anniversary.

Today, the Nanakuli–ʻEwa right-of-way is the longest remaining stretch of the historic OR&L. At the Railway Society Museum one can view Kauila No. 6 and No. 12 and equipment from the Oʻahu Sugar Company, Waialua Agricultural Company, and ʻEwa Plantation Company. Educational train rides conducted every Sunday at 1:00 and 3:00 PM cover approximately 11 miles roundtrip, and during the seventy-minute ride the railway's story and other local historical information is presented. Weekday group rides are available by reservation.

OLD HONOLULU POLICE STATION

During World War II, for nearly three years, provost marshal Col. William F. Steer was headquartered in the police station while Hawaiʻi was under martial law. At that time, the holding cell in the basement could accommodate up to 250 drunks, and on paydays and weekends there were roundups with fifteen to twenty trucks lined up along Ala Moana Boulevard. The station served as a police department headquarters from 1931 to 1961. Over the years the Honolulu Police Station handled 13 million telephone complaints, numerous tsunami warnings, and messages about the bombing of Pearl Harbor. From 1967 to 1982 it housed the state district court and traffic violations bureau.

The building was named the Walter Murray Gibson Building after Hawaiʻi's infamous "Minister of Everything." Walter Murray Gibson (1822–1888) rose to the height of political influence as a champion of the movement to restore power to the Hawaiian people, which had eroded under Western influence and political infiltration. He succeeded in alienating just about everyone in the government who was not Hawaiian. Gibson supported the teaching and use of the Hawaiian language at a time when others were trying to make English the standard language. He fought to prevent annexation to the United States and was constantly at odds with the growing majority of white businessmen and plantation owners.

When he was elected to the legislature in 1878, Gibson began a political career that brought him almost absolute control over all state affairs, as he became president of the board of health, prime minister, and minister of foreign affairs. He served as a substitute attorney general and temporarily took over the Department of the Interior.

At the end of Gibson's reign, opposition leaders attempted to expose him as a charlatan and opportunist. After barely escaping lynching, Gibson left the islands in exile in 1887 and died on January 21, 1888, at the age of sixty-six. His remains were shipped back to Hawaiʻi, and the *Hawaiian Gazette*, which had for years delighted in criticizing him, described the passage of his remains through customs as follows: "W. M. Gibson. One corpse. No value."

Five years after his death, the Hawaiian monarchy came to an end, and with this died the memory of Walter Murray Gibson. No street, lane, park, or building ever bore his name until recently. Even his gravesite is unknown. The man who had served Hawaiʻi and the Hawaiians most of his life, adopted their language and customs, supported the native populace against the outsiders, encouraged self-governance and monarchy rule, and been at the center of political power for over five years had disappeared without a trace. It was not until 1991 that tribute has been paid to Walter Murray Gibson in naming the old police station after him.

PUNAHOU SCHOOL

In 1849, during the period of land division known as the Great Mahele, the mission obtained a written title to the lands known as Ka Punahou. Queen Kaʻahumanu took special interest in the Punahou lands and in 1830 had her own thatched house built there. Near it she built a smaller house for the Binghams. She also erected a stone wall to protect the lands from grazing cattle; today the Kaʻahumanu Wall runs the length of the campus from Punahou to Clement Street, along Wilder Avenue.

By 1883 the Punahou buildings were so crowded that the trustees decided to install the Preparatory School in a building on Beretania Street, near the present St. Andrew's Cathedral. In 1884 the Bishop Hall of Science was built, and in 1885 a new home for the president was erected near Old School Hall. By 1898 a new school center was built by Charles Reed Bishop, who named it Pauahi Hall in memory of his wife, Bernice Pauahi Bishop. Cooke Hall was completed in 1908 and housed the school's library and an art gallery for the Cooke family collection and other donations. Many other buildings were added over the next several years, including Castle Hall, Montague Hall, Thurston Memorial Chapel, MacNeil Hall, Bingham Hall, Alexander Hall, Griffiths Hall, and Dillingham Hall.

World War II was an interesting and unusual time for Punahou School. At 1:10 AM on December 8, 1941, trucks of the U.S. Army Corps of Engineers moved onto the campus, occupied the buildings, and evacuated students. Barbed wire was set in place, and trenches were dug around the campus and lined with sod from Alexander Field. Cooke Library, containing the school's switchboard, became the military headquarters, and bomb-proof tunnels were dug between the library, Alexander Hall, and Pauahi Hall. The students rallied to support the war effort; boys took up guard duty, dug trenches, served as fire wardens and ambulance drivers, and worked in the pineapple fields and cannery to help relieve the labor shortage. Girls worked in the hospitals, making surgical dressings and assisting with patient care. Even the younger students joined in the massive fund-raising activities, selling nearly $5 million worth of war bonds and stamps. The money was used to purchase three bombers and one fighter plane and to equip a 2,000-bed hospital. After the war, Punahou was lauded for the unfailing cooperation and patriotism of its trustees, faculty, and students.

QUEEN EMMA SUMMER PALACE

Hanaiakamalama was a place for entertaining visiting dignitaries and friends of the royal family. An account of one such occasion appeared in the *Hawaiian Gazette*, March 2, 1870: "On Monday afternoon Her Majesty Queen Emma gave an impromptu entertainment to a large number of guests at her residence in Nu'uanu Valley. The guests enjoyed themselves at croquet and other outdoor sports on the lawn until evening when the fine room prepared for the entertainment of the Duke of Edinburgh was thrown open and dancing commenced and was kept up until about nine o'clock when they took leave." The room referred to is believed to have been a lanai at the rear of the house which had been converted to an extra room for a visit by the duke.

Displayed at the Queen Emma Summer Palace are many personal possessions and mementos of the royal family and their guests. The beautiful koa cradle made for Emma's only child, Albert Edward, is a poignant reminder of the beloved prince who died at age four; a locket bracelet containing a picture of Queen Victoria and a lock of her hair is testimony to the close friendship shared by the two queens. Queen Victoria was godmother to the Hawaiian prince, and for his christening she sent an ornate silver christening urn, also on display. A Gothic Revival cabinet made in Germany of Hawaiian koa wood for King Kamehameha IV and Queen Emma and a four-poster bed of koa are among the other royal possessions on view here.

THE ROYAL HAWAIIAN HOTEL

The Royal Hawaiian Hotel was one of three major Waikiki hotels of the post–World War I era designed to serve a burgeoning crowd of visitors from the U.S. mainland. The two other luxury hotels at the time were the Moana Hotel and the Alexander Young Hotel in downtown Honolulu. The Royal was intended to be Waikiki's major luxury hotel, supplanting the Moana, which had been built twenty-six years earlier.

The early 1920s was a period of great prosperity in Hawai'i and the United States. Business and government leaders in the islands saw increasing opportunity to develop a new type of visitor industry, referred to as the "carriage trade."

In 1922 the territorial government initiated the digging of the Ala Wai Canal, which made possible the development of Waikiki. The next step was to provide visitors with suitable transportation to the islands and luxurious accommodations once they arrived. This was achieved in 1925, when the Matson Navigation Company built the 650-passenger luxury liner *Malolo*—the fastest ship on the Pacific, with all the amenities one could desire. The ship's name was changed to the *Matsonia* and later to the *Lurline*.

The Territorial Hotels Company budgeted an unheard-of $2 million for the plush, 400-room Royal Hawaiian Hotel, to be the largest and most elegant in the islands. Many well-known people enjoyed their island sojourns at the Royal, including Franklin D. Roosevelt, Nelson Rockefeller, Shirley Temple, Jeanette MacDonald, Douglas Fairbanks, Jack Benny, Henry Ford II, Al Jolson, Mary Pickford, the Shah of Iran, and the Sheik of Kuwait.

During World War II, the U.S. Navy leased the Royal for a rest and recreation facility for the Pacific Fleet, charging officers one dollar a day and enlisted men nothing. After the war, the hotel saw extensive renovation and remodeling, and in 1969 the sixteen-story Royal Tower was added.

ST. ANDREW'S CATHEDRAL AND PRIORY

The St. Andrew's Priory School for Girls began in 1867, when the first buildings were completed. They were described as "quaint, one-story cloistered buildings . . . surrounded by a pretty court, whose marked feature was a tall cross of the Island coral." The goal of the priory sisters was "to direct the lives of the girls committed to its care on the highest and truest lines, fitting them for usefulness in whatever department of life they may be placed." In addition to the usual curriculum, the girls were taught to cook, sew, and do housework.

From its beginning, Saint Andrew's Cathedral and Priory has been the center of the Episcopal church and mission in Hawai'i. Today its diocese extends west to Okinawa and south to Samoa. The Priory School for Girls is an active private school situated next to the cathedral. A bust of Queen Emma in terra cotta, sculpted by Alice Campbell, is found in the small park adjacent to the school. Stories are told of the spirit of Queen Emma walking through the grounds, keeping a watchful eye on her beautiful school and cathedral. Saint Andrew's is her living memorial—a place of beauty and serenity in an area steeped in the history of the monarchy and the cultural influences of other lands.

STATUE OF KING KAMEHAMEHA I AND ALI'IOLANI HALE

The original statue of Kamehameha I was sent to Kohala, on the Big Island, the king's birthplace. On the day of its unveiling there, the concrete base had not yet dried, so the statue had to be hoisted over the base by a crane and suspended in a sling. When the veil was removed, the crowd gasped to see the sight of King Kamehameha the Great suspended as if from a gallows. The shock quickly passed and a great cheer arose. Guns fired a royal salute and the band played "Hawai'i Ponoi." Many onlookers placed wreaths and leis around the feet of the statue. It was the general consensus of Kohala residents that Kamehameha himself was behind the unusual set of circumstances that brought the statue to his hometown, where it rightfully belonged.

As a young man, King Kamehameha V had been to England and France and been impressed with the royal residences there. The royal house built by Kekuanao'a in 1845 for his daughter, Victoria, had grown rather decrepit. King Kamehameha intended to build Ali'iolani Hale as a symbol of the increasing prosperity of the independent Hawaiian kingdom.

Ali'iolani Hale was the scene of the deposition of Queen Lili'uokalani and the declaration of the provisional government in 1893. Events leading up to the overthrow of the monarchy were long in coming. Early in 1893 Queen Lili'uokalani had proposed a new constitution in reaction to the growing foreign involvement and annexationist influence in government affairs. A storm of protest arose among the annexationists, who formed the so-called Committee of Safety and took possession of the building on January 17, 1893. That afternoon, on the steps of Ali'iolani Hale, the proclamation was read which declared an end to monarchic rule in the Hawaiian Islands. This marked the beginning of government by foreign interests, particularly American. The U.S. flag was first unfurled in Hawai'i over Ali'iolani Hale, where the provisional government established itself. Later, the new executive and legislative offices were moved to 'Iolani Palace; Ali'iolani Hale became the Judiciary Building and eventually housed the territorial and then the state supreme and circuit courts.

THOMAS SQUARE

By the late 1830s and early 1940s, groups from several imperialistic countries had settled in Hawai'i. The kingdom was continuously troubled by the growing influence of foreigners and interference of foreign governments in its affairs. The tide of foreign influence was turning against sovereign rule, and Hawai'i's position as an independent nation was being threatened. During 1842, King Kamehameha III sent ambassadors to Great Britain, France, and the United States—three of the countries involved—in an attempt to gain recognition of Hawai'i's independence.

There were numerous claims to land ownership and desperate competition among the foreign-born residents to lay claim to more and more property. One such claimant was Englishman Richard Charlton, who not only insisted that land had been granted to him by Chief Boki but made several allegations of mistreatment of British subjects by the Hawaiian government. Charlton's charges led to the takeover of the government by Lord George Paulet in 1843. In his public speech to announce the forced cession, the king spoke these words in Hawaiian:

> Where are you, chiefs, people and commons from my ancestors and people from foreign lands! Hear ye, I make known to you that I am in perplexity by reason of difficulties into which I have been brought without cause; therefore I have given away the life of our land, hear ye! But my rule over you, my people, and your privileges will continue, for I have hope that the life of the land will be restored when my conduct shall be justified.

For five months the foreign rule under Paulet continued, supported by British naval guns and armed forces. The king fled to Lahaina, Maui, and stole back by night across the channel for secret meetings with his trusted friend and adviser, Dr. Gerrit Judd. Judd had secreted the government records in the royal crypt on the palace grounds and, using the coffin of Queen Ka'ahumanu as a desk, he carried on government correspondence and petitioned Great Britain to right the wrong that had been done by Lord Paulet.

U.S. POST OFFICE, CUSTOM HOUSE AND COURT HOUSE

The "Old Federal Building" maintains a place of prominence in the most historic section of downtown Honolulu as it continues to serve actively as a government building. It is noted for having the first female customs collector for the U.S. Customs Service and the only woman to occupy this post in Hawai'i. Jeanette Hyde was appointed by President Coolidge in 1925. One of her first actions as collector was to send 325 tins of opium to Washington as an example of customs efforts to stem the growing tide of narcotics smuggling in Hawai'i. During World War II, the building was a center for anti-espionage activities.

USS ARIZONA MEMORIAL

By April 1941 the Japanese had amassed several hundred planes in preparation for the attack on Pearl harbor. The summer months were spent preparing for the attack, perfecting dive-bombing maneuvers, and making short and shallow torpedo runs at remote sites. In early November, final approval was given and the date for the attack was chosen: December 7—a Sunday, when most ships would be in port for the weekend. Knowing that the Pacific Fleet would get under way in the event of an attack, the Japanese sent an advance expeditionary force of twenty-seven submarines ahead of the Pearl Harbor striking force. Their job was to send information to the strike force and torpedo any vessels that might escape the air attack. Five of these submarines carried battery-operated, midget two-man submarines, each carrying two torpedoes.

The Japanese succeeded in destroying most of the Pacific Fleet and wiping out airfields at Ford Island, Wheeler, Hickam, ʻEwa, Kaneʻohe, and Bellows to eliminate the threat of retaliation by air. Within minutes the majority of American fighters, bombers, and patrol planes were either destroyed or disabled. The Japanese striking force blasted its way to the primary targets, the eight battleships of the Pacific Fleet. Finding the ships moored in pairs, they attacked the inside ships with bombs and the outboard ships with torpedoes. The USS *Arizona* and USS *Vestal* were victims of this strategy. After completing the bombing runs, the attackers flew back over their targets, strafing them to kill any survivors.

The attack on Pearl Harbor lasted only about two hours, but it became the war cry of America that launched an incredible feat of reconstruction and retaliation. Hawaiʻi immediately became a war zone, with martial law imposed. All of Hawaiʻi joined the war effort, and the entire island of Oʻahu came to the support of the armed forces.

USS BOWFIN

Fleet-type submarines originated with the British Royal Navy during World War I when false rumors were being circulated about the Germans having produced such submarines. These boats were formidable weapons: they could take on the most powerful ships afloat with their conventional torpedoes, hit smaller, agile opponents with homing torpedoes, and use their deck guns on most lightly armed surface craft. They were extremely fast in tracking their targets, moving into position to fire, attacking, and diving.

After the USS *Bowfin*'s years of service, her fate would have been the scrap metal heap had the Pacific Fleet Submarine Memorial Association not intervened and restored her as a submarine memorial. On November 14, 1980, the Navy approved a five-year lease at the two-and-one-half acre Bowfin Park site, and in April 1981 the *Bowfin* was opened for visitors.

Through the efforts of many people and organizations, including the Dillingham Corporation, the Navy League, and scores of volunteers, the *Bowfin* was brought back into condition for exhibition. In 1984 the Pacific Fleet Memorial Association suggested that the submarine museum at Pearl Harbor Submarine Base be incorporated into Bowfin Park. The proposal was approved and the association negotiated a 25-year lease with the Navy in June 1986. During construction of the park, the *Bowfin* remained open for visitors and was even used in the TV production of "War and Remembrance." Bowfin Park was officially dedicated in September 1988. The USS *Bowfin* had already been declared a National Historic Landmark in January 1986.

Since its opening, Bowfin Park, together with its neighbor, the Arizona Memorial, has become one of the most popular attractions in Hawaiʻi. It provides a museum featuring the history of the submarine service, with related artifacts including uniforms, battle flags, photographs, artwork, a 10-foot model with cutaway views of the submarine, and a 25,000-volume research library. Bowfin Park hosts approximately 250,000 visitors a year.

WAR MEMORIAL NATATORIUM

The impetus to build a World War I memorial began shortly after the war years, when citizens expressed the need to pay tribute to the men and women who had served in the war. In 1919, Governor Charles McCarthy stated, "We owe it to the splendid young men of Hawaiʻi who went forth to the colors in answer to the call of their country, on land or on sea, to do something perpetuating the memory of their faith and allegiance."

The War Memorial Natatorium opened in 1927 with the following *Advertiser* commentary:

> Tonight the Hawai'i War Memorial opens. It is highly appropriate that this Memorial to the heroes of the World War should be a public natatorium. America went to war to assure safety and independence and the privileges and rights of a free people to all her citizens, and a part of the birthright of a free people is sound health and the opportunity for wholesome recreation. The Natatorium epitomizes Hawai'i's prominence in one of the world's great sports. Situated at Waikiki, it looks upon and is part of the ocean, whereof Hawai'i is the "crossroad."... No such galaxy of swimming stars has ever been gathered together since the last Olympic Games. The opening of the natatorium will be signalized by the greatest competitive swimming ever seen anywhere in the Pacific, once more giving Hawai'i a place of honor and distinction.

The colorful opening ceremonies included an AAU national championship swim meet, with swimmers from Japan and South America participating. Olympic champion Johnny Weissmuller broke the world's record for the 100-meter freestyle swim, and in the following three days of competition he set new world records for the 440- and 880-meter freestyles. Clarence "Buster" Crabbe, a local swimmer, won the 1500-meter contest—and later went on to replace Weissmuller in the famous "Tarzan" series.

WASHINGTON PLACE

The two years of Queen Lili'uokalani's reign was a period of political turmoil. In reaction to the infamous Bayonet Constitution forced upon her brother, King Kalakaua, in 1887, the queen was determined to assert her sovereignty and promote the rights of her native people. But opposition from the legislature, economic conditions, and the queen's own inexperience at government were against her.

Under the McKinley Tariff of 1890, Hawaiian sugar lost its advantage of free and favored entry into the United States. The resulting economic depression served only to weaken the queen's position and strengthen the opposition. In January 1893, Queen Lili'uokalani dissolved the existing legislature and attempted to promulgate a new constitution. A coup led by the self-appointed Committee of Safety moved to overthrow her and set up a provisional government. They forced Queen Lili'uokalani to yield control on January 17, 1893. A counter-revolution of Hawaiian supporters in 1895 proved disastrous. After a small arsenal of arms and dynamite was discovered at Washington Place, the queen was arrested and imprisoned for eight months in 'Iolani Palace. Upon her release, Queen Lili'uokalani returned to her home at Washington Place, where she lived for nearly two decades until her death in 1917 at age seventy-nine.

IBLIOGRAPHY

Adler, Jacob, and Robert M. Kamens. *The Fantastic Life of Walter Murray Gibson, Hawaii's Minister of Everything.* University of Hawaii Press. Honolulu, 1986.

Allen, Gwenfread E. *Hawaii's Iolani Palace and Its Kings and Queens.* Aloha Graphics. Honolulu, 1978.

Allen, Helena G. *The Betrayal of Liliuokalani: Last Queen of Hawaii.* Mutual Publishing. Honolulu, HI, 1982.

American Institute of Architects. Hawaii Society. *Oral Histories of the 1930's Architects: Transcriptions of Tapes of Oral Histories.* Hawaii State Historic Preservation Office. Honolulu, 1982.

Andrews, R. W. *The Story of Three Old Buildings in Honolulu.* Hawaii Mission Children's Society. Honolulu, 1926.

Bailey, Paul. *Those Kings and Queens of Old Hawaii.* Westernlore Books. Los Angeles, 1975.

Bartlett, Rev. S. C. *Historical Sketch of the Missions of the American Board in the Sandwich Islands, Micronesia, and Marquesas.* American Board of Commissioners for Foreign Missions. Boston, 1880.

Belknap, J. P. *Majesty: The Exceptional Trees of Hawaii.* Outdoor Circle. Honolulu, 1982.

Bell, Roger J. *Last among Equals: Hawaiian Statehood and American Politics.* University of Hawaii Press. Honolulu, 1984.

Board of Trustees. *Excerpts from the will and Codicils of Princess Bernice Pauahi Bishop.* Bishop Museum. Honolulu, 1976.

Bradley, Harold. *The American Frontier in Hawaii— The Pioneers: 1789–1843.* Stanford University Press. Stanford, Calif., 1942.

Bushnell, Oswald A. *A Walk through Old Honolulu.* Kapa Associates. Honolulu, 1975.

Carter, G. K. *Sightseeing Historic Honolulu.* Tongg Publishing. Honolulu, 1956.

Conrad, Agnes, and Barbara Dunn. *Hawaii Museums and Related Organizations.* Hawaii Museums Association and Hawaii Historical Society. Honolulu, 1988.

Daws, Gavan. *Shoal of Time: A History of the Hawaiian Islands.* University of Hawaii Press. Honolulu, 1968.

Day, A. Grove. *Hawaii: Fiftieth Star.* Sloan and Pearce. New York, 1960.

Day, A. Grove. *History Makers of Hawaii.* Mutual Publishing. Honolulu, HI, 1984.

Day, A. Grove. *Kamehameha, First King of Hawaii.* Hogarth Press. Honolulu, 1974.

Dean, Love. *The Lighthouses of Hawaii.* University of Hawaii Press. Honolulu, 1991.

Dodge, Charlotte Peabody. *Punahou: The War Years.* Punahou School. Honolulu, 1984.

Farrell, Andrew. *The Story of Iolani Palace.* Board of Commissioners of Public Archives. Advertiser Publishing. Honolulu, 1936.

Fitzgerald, Donald T. *A History of the Kakaako Pumping Station.* Historic Hawai'i Foundation. Honolulu, 1992.

Free, David. *Hawaii's Builders and Dreamers.* Pacific Business News. Honolulu, 1984.

Gessler, C. *Hawaii: Isles of Enchantment.* D. Appleton Century. New York, 1937.

Gessler, C. *Tropic Landfall: The Port of Honolulu.* Doubleday, Doran. New York, 1942.

Hackler, Rhoda E. *'Iolani Palace.* Friends of 'Iolani Palace. Honolulu, 1987.

Hearst, Anne Randolph. *Celebrating a Palace Evolution.* Town & Country Magazine, 1987.

Hibbard, Don, and David Franzen. *The View from Diamond Head: Royal Residence to Urban Resort.* Editions Limited. Honolulu, 1986.

Hawaiian Journal of History. "The Kamehameha Statue." Vol. 3 (1969).

Historic Hawaii News.. "Academy of Arts." November/December 1982, p.3.

"Alexander & Baldwin Building. " August 1986, p. 17.

"Aliiolani Hale." February 1987, pp. 8-9; June 1980, p. 3.

"Aloha Tower." May 1987, p. 13.

"Architecture of C. W. Dickey." December 1988, pp. 4-6.

"Armed Services YMCA." December 1988, p. 9; November 1989, p.10.

"Bishop Hall." December 1989, p. 9.

"Cathedral of Our Lady of Peace." July/August 1992, pp. 4-7; January 1979, p. 1.

"Cooke, Anna Charlotte Rice." February 1988, p. 16
"Coronation Pavilion." May 1981, p. 8.
"Diamond Head Lighthouse." October 1986, p. 22.
"Dillingham Transportation Building." May 1980, pp. 6-8.
"Ewa Plantation." March 1988, pp. 7-17.
"Falls of Clyde." August 1988, p. 4; June 1985, p. 5; June 1989, p. 11.
"Hanaikamalama." August 1989, pp. 6-7.
"Hawaiian Electric Company." February 1987, pp. 4-6.
"Honolulu Academy of Arts." February 1988, p. 16.
"Honolulu Hale." May 1987, p. 10.
"Honolulu Police Station, Old." May 1987, p. 15.
"Iolani Barracks." January 1987, p. 5; May 1981, p. 9.
"Iolani Palace." May 1981, p. 8.
"James Campbell." March 1988, pp. 4-5.
"Kamehameha V Post Office." August 1988, p. 16; May 1987, p. 14.
"Kamehameha V Post Office Park." February 1976, pp. 4-5, 8.
"Kamehameha Schools." November 1986, pp. 10-11.
"La Pietra." January/February 1977, p. 4; May 1988, p. 6.
"Linekona School." July 1986, p. 17.
"Linekona School & McKinley High School." October 1990, pp. 10-11.
"Maintaining Historicity: Treasures of Honolulu That Keep the Fabric Intact." May 1987, pp. 13-15.
"Moana Hotel." July 1989, pp. 4-9; February 1990, p. 8; August 1988, pp. 8-9; August 1987, pp. 4-5.
"Oahu Railway and Land Company." January 1985, p. 2; March 1988, p. 7; February 1985, p. 1.
"Queen Emma Summer Palace." August 1989, pp. 4-7.
"Royal Brewery." December 1984, p. 1.
"Royal Hawaiian Hotel." November 1988, pp. 8-9.
"Royal Mausoleum." May 1986, p. 10; November 1986, pp.12-13; June 1987, p. 5.
"US Post Office/Custom House/Court House." February 1987, p. 4.
"USS Arizona Memorial." August 1989, p. 10; October 1989, p. 7.
"USS Bowfin." May 1986, p. 22.
"Waikiki War Memorial and Natatorium." January 1987, pp. 4-5.

Holt, John Dominis. *Monarchy in Hawaii.* Hogarth Press. Honolulu, 1971.

Honolulu Advertiser. "Friends Are Gathering to Save Ewa's Heritage." June 28, 1992.
"Friends for Ewa Group Want to Preserve Historic Plantation Character." April 30, 1989.
"Historic Hawaii Salutes 10 Historic Preservation Examples." January 20, 1991.
"Historic Hawaii to Present 9 Awards for Preservation." November 2, 1992.
"Historic Preservation Group Interested in Preserving Ewa Village." March 31, 1991.
"Historic Preservation Planned for Kakaako Pumping Station on Ala Moana." November 10, 1991.
"Journal of Hawaiian History Is a Historic Achievement." February 11, 1989.
"The Moana Hotel." March 3, 1989.
"Natatorium: What to Do?" October 30, 1981.

Honolulu Magazine. "To Share What She Had Seen: Honolulu Academy of Arts and Its Founder." April 1967, pp. 27, 41.

Honolulu Star-Bulletin. "Ewa Plantation Camps on Way to Renovation." October 5, 1990.
"Natatorium Ceremony Boosts Restoration." November 12, 1990.
"The Old Honolulu Police Station." February 13, 1985.

Jackson, Frances, Agnes Conrad, and Nancy Bannick. *Old Honolulu: A Guide to Oahu's Historic Buildings.* Historic Buildings Task Force. Honolulu, 1969.

Jay, Robert. *The Architecture of Charles W. Dickey.* University of Hawaii Press. Honolulu, 1992.

Joesting, E.. *Hawaii: An Uncommon History.* Norton. New York, 1972.

Jones, Maude. "Thomas Square." State of Hawaii Archives Historical File. Honolulu.

Judd, Walter F.. *Palaces and Forts of The Hawaiian Kingdom.* Pacific Books. Palo Alto, Calif., 1975 .

Kimmett, L., and M. Regis. *The Attack on Pearl Harbor: An Illustrated History.* Navigator Publishing. Seattle, 1991.

Kuykendall, R. *Hawaii in the World War.* Publications of the Historical Commission of the Territory of Hawaii, Vol. 2 (1928).

The Hawaiian Kingdom. 3 vols. University of Hawaii Press. Honolulu, 1957, 1966–67. and A. Grove Day. *Hawaii: A History.* Prentice-Hall. Englewood Cliffs. N. J., 1976.

Lili'uokalani. *Hawaii's Story by Hawaii's Queen.* Charles E. Tuttle. Rutland, Vt., and Tokyo, 1964.

Lott, A., and R. Sumrall. Ship's Data No. 5 *USS Bowfin (SS 287).* Leeward Publications. Annapolis, M.D., 1975.

Mellen, Kathleen. *An Island Kingdom Passes; Hawaii Becomes American.* Hastings House. New York, 1958.

Mid-Pacific Magazine. "The Bishop Museum." Vol. 7 (January 1914), pp. 94-96.

Mrantz, Maxinne. *Women of Old Hawaii.* Tongg Publishing. Honolulu, 1975.

Myatt, Carl. *Hawaii: The Electric Century.* Signature Publishing. Honolulu, 1991.

Neil, J. Meredith. *Paradise Improved: Environmental Design in Hawaii.* American Association of Architecture Bibliographers, Papers, Vol. 8 (1971).

Nelson, Paul A. *Bowfin Park: A Dream Becomes a Reality.* Honolulu Council, U.S. Navy League, 1989.

Paradise of the Pacific. "Aloha Tower: The Liberty Statue of the Pacific." Vol. 70 (March 1958), pp. 18-21.
"Character Building in the Hawaiian Islands: Spirit of Missions." Vol. 74 (July 1909), pp. 578-580.
"The Churches of Hawaii." Vol. 74 (December 1962), pp. 34-35.
"Halekoa, The End of an Epoch." Vol. 75 (April 1963), pp. 14-16, 30-31.
"Halekulani, House Befitting Heaven." Vol. 64 (September 1952), pp. 34-35.
"Historic Washington Place." Vol. 47 (1955 annual ed.), pp. 6-8, 116.
"History of Iolani Palace." Vol. 43 (November 1930), pp. 9-12.
"History of Kawaiaha'o Church." Vol. 5 (February 1892), p. 6.
"Honolulu Academy of Arts." Vol. 40 (May 1927), pp. 16-18.
"Royal Hawaiian Hotel Regains Glamour." Vol. 59 (February 1947), pp. 17-20.
"Royal Tombs of Hawaii." Vol. 41 (June 1928), pp. 13-15.
"St. Andrew's Priory." Vol. 23 (December 1910), pp. 59-61.
"The Story of Kawaiaha'o Church." Vol. 14 (January 1901).
"Two Public Buildings." Vol. 6 (July 1893), pp. 97-101.
"Westminster of Hawaii." Vol. 74 (December 1962).

Peterson, Barbara, ed. *Notable Women of Hawaii.* University of Hawaii Press. Honolulu, 1984.

Peterson, Charles E.. "American Notes: The Iolani Palaces and the Barracks." *Journal of the Society of Architectural Historians,* Vol. 22 (May 1963).

"Pioneer Architects and Builders of Honolulu." *Hawaii Architect,* November, 1978.

Pratt, C. Dudley. *HEI—The Start of a New Hawaii Tradition: An Electric Utility's Evolution from the Monarchy to the 21st Century.* The Newcomen Society of the United States. New York, 1988.

Rice, William T. *Pearl Harbor Story.* Tongg Publishing. Honolulu, 1965.

Rose, Roger G. *A Museum to Instruct and Delight: William T. Brigham and the Founding of the Bernice Pauahi Bishop Museum.* Bishop Museum Press. Honolulu, 1980.

Schoofs, Robert. *Pioneers of the Faith: History of the Catholic Mission in Hawaii, 1827–1940.* Boeynaems. Waikane, Hawaii, 1978.

Sheehan, Edward. *Day of '41: Pearl Harbor Remembered.* Kapaa Associates. Honolulu, 1976.

Sister Grace Marian. *The Honolulu Academy of Arts: Its Origin and Founder.* Honolulu Academy of Arts, 1984.

Sterling, Elspeth P., and C. Summer. *Sites of Oahu.* Bishop Museum. Honolulu, 1978.

Stilwell, P., ed. *Air Raid: Pearl Harbor: Recollections of a Day of Infamy.* Naval Institute Press. Annapolis, Md., 1981.

Stone, Scott S. C. *Pearl Harbor: The Way It Was.* Island Heritage. Honolulu, 1977.

Straus, Leon. *The Honolulu Police Department: A Brief History.* The 200 Club. Honolulu, 1978.

Sullivan, Josephine. *A History of C. Brewer Co., Ltd.: One Hundred Years in the Hawaiian Islands.* Walton Advertising & Printing. Boston, 1926.

Swenson, J., and E. Midkiff. *Treasures of the Hawaiian Kingdom.* Daughters of Hawaii. Honolulu, 1984.

Tabrah, Ruth. *Hawaii, Bicentennial History.* W. W. Norton. New York, 1980.

The Kamehameha Schools/Bernice Pauahi Bishop Estate. Fouders Day Celebration, *Sesquicentennial Souvenir Booklet.* December 1981

Taylor, Albert Pierce. *The Rulers of Hawaii.* Advertiser Publishing. Honolulu, 1927.

Thrum's Annual. "Hawaiian War Memorial." 1920.

"The Moana Hotel: Waikiki's New Attraction." 1901, pp. 161-165.

Jeannette Murray Peek's earliest memories were of drawing pictures of horses, animals, and scenery. Born in Troy, N.Y., she grew up in a family of eight children on an upstate New York farm. At age eighteen, she set off for Europe and lived in Paris, France, for one year where she frequented art galleries and filled her sketchpad with drawings of the city. She returned to America and pursued a bachelor's degree in Psychology and a master's degree in Rehabilitation Counseling from the State University of New York at Albany. Jeannette received no formal training in art but pursued drawing and painting as an avocation. She taught adult education art classes in California and New York. Her work has been exhibited in art shows, shops, and galleries in California and Hawai'i. Since coming to Hawai'i in 1979, her drawings of historic landmarks have gained her a reputation as one of Hawai'i's foremost pen and ink artists. In addition to drawing historic sites, she enjoys doing etchings and landscapes in watercolor and pastels. Jeannette Murray Peek's drawings have found their way into homes around the world. In 1989, her drawing of the Vallejo Naval and Historical Museum in California was chosen as the gift for Vallejo's sister city in Italy. Jeannette lived in Hawai'i from 1979 to 1994. She was employed as a counselor at Honolulu Community College. Today, she and her husband, Joseph, reside in North Carolina.